Replicated Data Management for Mobile Computing

Synthesis Lectures on Mobile and Pervasive Computing

Editor

Mahadev Satyanarayanan, *Carnegie Mellon University*

Mobile computing and pervasive computing represent major evolutionary steps in distributed systems, a line of research and development that dates back to the mid-1970s. Although many basic principles of distributed system design continue to apply, four key constraints of mobility have forced the development of specialized techniques. These include unpredictable variation in network quality, lowered trust and robustness of mobile elements, limitations on local resources imposed by weight and size constraints, and concern for battery power consumption. Beyond mobile computing lies pervasive (or ubiquitous) computing, whose essence is the creation of environments saturated with computing and communication yet gracefully integrated with human users. A rich collection of topics lies at the intersections of mobile and pervasive computing with many other areas of computer science.

RFID Explained
Roy Want
2006

Controlling Energy Demand in Mobile Computing Systems
Carla Schlatter Ellis
2007

Application Design for Wearable Computing
Dan Siewiorek, Asim Smailagic, and Thad Starner
2008

Location Systems: An Introduction to the Technology Behind Location Awareness
Anthony LaMarca and Eyal de Lara
2008

Replicated Data Management for Mobile Computing
Douglas B. Terry
2008

Replicated Data Management for Mobile Computing
Douglas B. Terry
www.morganclaypool.com

ISBN: 9781598292022   paperback

ISBN: 9781598292039   ebook

DOI: 10.2200/S00132ED1V01Y200807MPC005

A Publication in the Morgan & Claypool Publishers series

SYNTHESIS LECTURES ON MOBILE AND PERVASIVE COMPUTING #5

Lecture #5

Series Editor: Mahadev Satyanarayanan, Carnegie Mellon University

Series ISSN
ISSN 1933-9011   print
ISSN 1933-902X   electronic

Replicated Data Management for Mobile Computing

Douglas B. Terry
Microsoft Research

SYNTHESIS LECTURES ON MOBILE AND PERVASIVE COMPUTING #5

MORGAN & CLAYPOOL PUBLISHERS

ABSTRACT

Managing data in a mobile computing environment invariably involves caching or replication. In many cases, a mobile device has access only to data that is stored locally, and much of that data arrives via replication from other devices, PCs, and services. Given portable devices with limited resources, weak or intermittent connectivity, and security vulnerabilities, data replication serves to increase availability, reduce communication costs, foster sharing, and enhance survivability of critical information. Mobile systems have employed a variety of distributed architectures from client–server caching to peer-to-peer replication. Such systems generally provide weak consistency models in which read and update operations can be performed at any replica without coordination with other devices. The design of a replication protocol then centers on issues of how to record, propagate, order, and filter updates. Some protocols utilize operation logs, whereas others replicate state. Systems might provide best-effort delivery, using gossip protocols or multicast, or guarantee eventual consistency for arbitrary communication patterns, using recently developed pairwise, knowledge-driven protocols. Additionally, systems must detect and resolve the conflicts that arise from concurrent updates using techniques ranging from version vectors to read–write dependency checks. This lecture explores the choices faced in designing a replication protocol, with particular emphasis on meeting the needs of mobile applications. It presents the inherent trade-offs and implicit assumptions in alternative designs. The discussion is grounded by including case studies of research and commercial systems including Coda, Ficus, Bayou, Sybase's iAnywhere, and Microsoft's Sync Framework.

KEYWORDS

mobile data management, replication, caching, mobility, ubiquitous computing, disconnected operation, intermittent connectivity, data consistency, conflict detection, update propagation, epidemic algorithms

Contents

List of Figures

CHAPTER 1

Introduction

Mobility has become increasingly important for both business and casual users of computing technology. With the widespread adoption of portable computing devices, such as laptops, PDAs, tablet computers, music players, and cell phones, people can have almost constant access to their personal data as well as to information that is shared with others. A user drinking coffee in a cybercafé in India can access e-mail residing on a mail server in Seattle. A doctor in New York can monitor the health of patients in remote parts of Africa. A mother waiting to pick up her children after school can be instantly notified that her daughter's soccer practice has been moved to a new location. Teenagers congregating at the mall can use their cell phones to locate not only their buddies but also the hottest sales. Advances in wireless technology, such as WiFi and WiMax, allow people to communicate from their computers with friends, colleagues, and services located around the world. However, providing users anytime, anywhere access to contextually relevant information presents substantial challenges to designers of mobile computing systems.

Compared with distributed systems in which powerful computers are connected over a fixed networking infrastructure, such as the ubiquitous Internet, mobile computing environments differ in a number of fundamental ways. Specifically, mobile computing systems must accommodate three novel aspects:

1. *Portable devices with limited displays, CPU resources, storage, battery life, and security:* With improvements in flash memories and digital media cards, handheld devices can hold gigabytes of information, making them capable of storing much of a person's personal information such as appointments, addresses, e-mail, and even a substantial music collection. Similarly, faster CPUs make these devices capable of running a broad spectrum of both data-intensive and computing-intensive applications. However, portable devices will always lag behind desktop computers and server machines in computing and storage capacity. Moreover, the physical size of such devices places a premium on screen space and prohibits conventional input capabilities such as keyboards. Battery life will remain a critical resource. Since computation and communication both consume energy, mobile applications must conserve battery life by limiting all forms of resource usage. Furthermore, mobile devices are less secure and less robust than server or desktop machines. In particular,

they are routinely lost, damaged, or stolen, thereby demanding new approaches to information privacy and preservation.

2. *Intermittent, low-bandwidth, high-latency network connections:* Although wireless networks, such as metropolitan area WiFi networks, are being deployed at a rapid pace, ubiquitous connectivity for mobile devices remains an elusive goal. For instance, although the WiMax standard for broadband wireless access was recently approved, practical deployments are not expected until late 2008 at the earliest. The good news is that new cell phones and PDAs are shipping with Bluetooth networking, and near-field communication technology is being integrated into some devices, making them capable of direct communication with nearby neighbors. Such devices allow people in the same vicinity to directly share information including music, photos, and schedules. Wireless mesh networks can be established for communicating over larger distances. However, such connections are temporary at best, since they break when the device owners move to new locations. Additionally, communication delays can be high because of long network discovery times and multiple hops. Thus, mobile systems must accommodate intermittent and generally low-bandwidth, high-latency wireless connectivity between devices while being able to exploit higher bandwidth, more robust communications when the opportunity arises.

3. *Changing environmental conditions and contexts:* A person's context, including his location, time of day, schedule, colleagues, deadlines, and interests, affects his information needs and hence the service desired from mobile applications. Such applications must therefore adapt their behavior to each user's changing context as well as to changes in the available computing resources, including network bandwidth and remaining battery life. An application's adaptability can have a drastic impact on its overall usability. Location-aware data delivery, for example, can tailor the information communicated to mobile users based on a user's current location or intended destination. Similarly, social networks constructed from buddy lists are gaining popularity as a means of sharing information, but such networks often experience rapid flux.

Techniques that have been developed specifically for mobile computing systems include replication and caching of data for off-line access, remotely accessing data that resides on other machines, offloading computation onto servers or surrogate PCs, and adapting system policies and mechanisms to users' changing context and hence changing information needs. This lecture focuses primarily on the first of these issues, replication, although adaptation also plays a role. Replication among networked services, stationary PCs, and mobile devices serves to reduce access costs, increase availability, and enhance survivability for personal, public, and enterprise information. In other words, replicated data management will remain an essential aspect of mobile computing for the foreseeable future.

1.1   HISTORICAL PERSPECTIVE

Mobile computing is a recent phenomenon, having taken off in the past 10 to 15 years. However, the fundamental technologies used in mobile computing systems were developed starting in the late 1960s and 1970s. From a hardware viewpoint, the 1970s saw the invention of PCs, such as the Xerox Alto, and wireless communication networks, such as the ALOHA network; the Dynabook concept, a vision of an intensely personal mobile computing device, inspired researchers for decades to come. In terms of software, this is when personal computing environments were developed along with the first replication protocols.

The key concepts presented in this lecture started emerging in the early 1980s, although initially without mobility as a focus. Table 1.1 summarizes many of the key technology innovations that relate to replicated data management for mobile computing, together with significant commercial and research systems. To provide some context for these developments, this table also includes major hardware and platform milestones. The dates shown indicate when papers on the systems were first published, when technical specifications were made available, or when commercial products hit the market.

The Grapevine system, developed at Xerox PARC and first published in 1981 [9], was a research prototype providing a replicated directory and e-mail service. It was deployed on a couple of dozen geographically distributed servers and widely used throughout Xerox. Perhaps most significantly, although previous replication protocols focused on maintaining mutually consistent replicated databases, Grapevine demonstrated the benefits of weak consistency replication and showed that applications could be designed to tolerate temporarily inconsistent data access, a hallmark of virtually all mobile systems today. This work not only led to a commercial product, called the Xerox Clearinghouse, which was the first system to propagate data among replicas using epidemic algorithms [12], but also planted the seeds for the Bayou project a decade later.

Locus was a distributed operating system developed at UCLA starting in 1979 that, among other novel features, included a replicated, network file system [100]. Notably, Locus allowed users of stationary computers to continue to access files despite failed servers and network partitions. Version vectors, developed to detect concurrent file updates [64], have been widely adopted for conflict detection in mobile systems (see Section 6.2.4). A sequence of follow-on research projects at UCLA, extending through the 1990s, explored replicated file systems for mobile users. These included Ficus [24, 73], one of the first mobile file systems with a peer-to-peer replication model and automatic conflict resolution, and Roam [74], which increased the scalability of Ficus by grouping devices into "wards."

The mobile computing community has also benefited from techniques for managing concurrent updates that were developed for collaboration systems, also known as groupware, starting in the mid-1980s. Such systems often allow users to operate on shared data in parallel while merging and resolving conflicting operations after the fact. For example, a group editor called Grove pioneered

TABLE 1.1: Timeline of significant events in mobile computing

YEAR	SYSTEM/ TECHNOLOGY	SIGNIFICANCE
1981	Grapevine	Showed that practical systems could use weak consistency replication
1983	Locus	Devised version vectors for conflict detection
1987	Clearinghouse	Commercial product relying on epidemic algorithms for update propagation
1987	Laptops	Provided a truly mobile platform for serious computing
1989	Grove	Group editor using operation transformation
1989	Lotus Notes	Commercial product for document replication via periodic bidirectional data exchanges
1990	Coda	First distributed file system to support disconnected operation; later explored weakly connected operation and automatic conflict resolution
1991	GSM	Second-generation cellular telephone network launched in Finland
1991	Ubiquitous computing	Vision for mobile computing pioneered at Xerox PARC
1993	Apple Newton	First commercial PDA
1993	Ficus	Peer-to-peer replicated file system with conflict resolution
1994	Bayou	Replicated database with application-specific conflict management and session guarantees
1994	Bluetooth	Industry standard short-range wireless protocol developed, although devices did not hit the market until several years later
1996	Palm Pilot	First widely adopted PDA with sync capability
1997	WiFi	High-speed wireless local-area networking standard
1997	WAP	Forum established to standardize wireless Web access
1998	Roam	Introduced ward model for scalable peer-to-peer replication
1999	BlackBerry	Commercial cell phone popularizing mobile e-mail access
2001	IceCube	Allowed application-provided ordering constraints on operations
2002	TACT	Explored alternative, bounded consistency models
2004	PRACTI	Separated invalidation notifications from updates in a log-based replication system

the notion of operation transformation [17], which permits operations from different users (or different mobile devices) to be applied in whatever order they arrive while guaranteeing that replicas reach a mutually consistent state (see Section 4.4.6). The Lotus Notes system, also targeted at collaborative applications, further validated the acceptance of weak consistency replication and demonstrated peer-to-peer synchronization for document databases.

Many innovations in the mobile computing space came out of the Coda project at Carnegie Mellon University (CMU), which started in 1987 and is still going strong [86]. The project's initial focus was on extending the Andrew File System (AFS) to provide increased fault tolerance through replication of file volumes across servers. Mobility was not an issue because laptops were just entering the market in earnest. However, Coda's client–server model with client-side caching proved attractive to the rapidly growing community of laptop users in the early 1990s. Thus, Coda was well-positioned to be at the forefront of mobile computing research. The project gained widespread recognition for its development of disconnected operation and followed up with advances in file hoarding, trickle reintegration, automatic conflict resolution, and a variety of other file service enhancements intended to better support mobile users, many of which are reported on in later sections.

Around 1990, researchers at Xerox PARC articulated a vision, called "ubiquitous computing," in which varied devices situated in the environment seamlessly interact with mobile users to provide continuous access to data and other computing resources [107]. Shortly thereafter, the Bayou project emerged with the goal of designing a data management platform in support of ubiquitous computing applications [13]. Bayou, in contrast to Coda, explored a peer-to-peer replication model for relational databases. Unlike previous systems that stressed replication transparency, Bayou was novel in its support for application-specific communication patterns, conflict management, and consistency guarantees.

In the past decade, along with an upsurge of mobile devices and wireless networks, mobile computing research has taken off with yearly conferences presenting significant advances and fresh applications. Systems such as IceCube [43], TACT [104], and PRACTI [8] have extended the state of the art in mobile data management by building on previous work. The contributions of these more recent projects are sprinkled throughout this lecture.

1.2 LECTURE ORGANIZATION

This lecture covers the basic techniques used for managing data replicas across mobile devices. It is organized as follows:

- Chapter 2 presents the basic terminology used throughout this lecture and describes various system models involving mobile devices, including commonly used client–server and peer-to-peer architectures. Replication requirements are extracted from these models and data usage patterns.

- Chapter 3 defines a variety of consistency guarantees that may be provided by a replicated system. Within the class of weak consistency protocols suitable for intermittently connected devices, "eventual consistency" is the most popular property, but others are possible, ranging from best effort convergence to bounded inconsistency and hybrid schemes.

- Chapter 4 provides the core lecture material, namely, the detailed design and implementation of a broad spectrum of replication protocols. Issues covered include representing, tracking, propagating, and ordering updates. The trade-offs and assumptions behind different approaches are discussed.

- Chapter 5 extends the protocols discussed in Chapter 4 to allow partial replicas that store select items from a large data collection.

- Chapter 6 focuses on an important issue faced by weak consistency replication protocols: how to detect and resolve conflicting updates made concurrently on different devices.

- Chapter 7 presents several research and commercial systems as case studies, referring back to the main techniques presented in earlier chapters.

- Chapter 8 concludes with general observations about mobile data management.

The bibliography contains a compilation of publications on topics related to this lecture. Throughout this lecture, citations are included to these published papers from the mobile computing literature. The interested reader should consult these papers for additional material.

* * * *

CHAPTER 2

System Models

Mobile systems have used a number of different models for how data is accessed, where it is stored, who is allowed to update it, how updated data propagates between devices, and what consistency is maintained. This chapter explores some common alternative models and concludes with the requirements they place on replication protocols. Think of these as choices that one faces when designing a mobile system. In practice, some systems fall squarely in one design space, whereas others are a hybrid with a mixture of system models. Before describing the models, the next section introduces the basic components that are common to all systems and the terminology used throughout this lecture.

2.1 BASIC COMPONENTS AND TERMINOLOGY

A mobile system comprises a number of *devices* with computing, storage, and communication capabilities. A wide variety of *mobile devices* may participate in a system, including laptops, tablet PCs, PDAs, cell phones, music players, video players, navigation systems, portable game players, digital cameras, electronic photo frames, health monitors, and smart watches. All such devices are capable of storing and communicating significant amounts of information. Mobile devices also include computing, storage, and communication components that are attached to moving vehicles, such as buses and trains, and even to animals. Mobile devices may interoperate not only with other mobile devices but also with *stationary devices*, such as desktop PCs and server machines.

Devices can communicate with each other over a spectrum of networking technologies. Most devices come equipped with an Ethernet port for connecting to *wired networks* such as the Internet as well as USB ports for connecting directly to other devices. Mobile devices also frequently, but not always, have one or more *wireless networks*, such as WiFi (i.e., IEEE 802.11), Bluetooth, and/or cellular.

Two devices are *connected* if they can send messages to each other, either over a wireless or wired network. *Weakly connected* devices can communicate, but only using a low-bandwidth, high-latency connection. A device is said to be *disconnected* if it cannot currently communicate with any other device. In practice, a given device may be connected to some devices and disconnected from others. Devices may experience *intermittent connectivity* characterized by alternating periods in which they are connected and disconnected.

An *item* is the basic unit of information that can be managed, manipulated, and replicated by devices. Items include photos, songs, playlists, appointments, e-mail messages, files, videos, contacts, tasks, documents, and any other data objects of interest to mobile users. Some items, such as digital photos, may exist only in the electronic world, whereas others, such as address book entries, may contain information about physical objects. Each item can be named by some form of *globally unique identifier*.

A *collection* is a set of related items, generally of the same type and belonging to the same person. For example, "Joe's e-mail" is a collection of e-mail messages, "Mary's calendar" is a collection of appointments, and "Suzy's picture gallery" is a collection of digital photos. A collection is an abstract entity that is not tied to any particular device or location or physical storage representation. Like items, each collection has a globally unique identifier so devices can refer to specific collections in replication protocols.

Collections can be shared and replicated among devices. A *replica* is a copy of items from a collection that is stored on a given device. A replica is a *full replica* if it contains all of the items in a collection. As new items are added to a collection, copies of these items automatically appear in every full replica of the collection. A *partial replica* contains a subset of the items in a collection. Devices maintain their replicas in local, persistent storage, called *data stores*, so that the replicated items survive device crashes and restarts.

Software applications running on a device can access the device's locally stored replicas and possibly replicas residing on other connected devices. Such applications can perform four basic classes of operations on a replica:

> A *read* operation returns the contents of one or more items from a replica. Read operations include retrieving an item by its globally unique identifier, as in a conventional file system read operation, as well as querying items by content.
>
> A *create* operation generates a new item with fresh contents and adds it to a collection. This item is first created in the replica on which the create operation is performed, usually the device's local replica, but is then replicated to all other replicas for the same collection.
>
> A *modify* operation changes the contents of an item (or set of items) in a replica, producing a new version of that item. A file system write operation is an example of one that modifies an item. A SQL update statement on a relational database is also a modify operation.
>
> A *delete* operation directly removes an item from a replica and the associated collection. Because the item is permanently deleted from its collection, it will be removed from all replicas of that collection. By contrast, a device holding a partial replica may choose to *discard* an item from its replica to save space without causing that item to be deleted from the collection.

An *update* is a generic term for a create, modify, or delete operation. The replication protocols discussed later are mainly concerned with propagating updates between replicas. When an update is made directly to an item in a device's replica, that device is said to have *updated* the item. Not all operations can necessarily be performed on all replicas. For instance, a *read-only replica* residing on a device might allow read operations but prevent update operations. In some of the system models discussed below, items are created but never modified. In this case, replicas contain *read-only items*.

2.2 REMOTE DATA ACCESS

Perhaps the most basic model for providing anytime, anywhere access to shared information is to store such information on a server machine from which it can be remotely fetched by mobile and wireless devices. For example, as depicted in Figure 2.1, a user can browse the Web from his cell phone using the standard wireless access protocol (WAP). Other examples include accessing large databases from a laptop or PDA using WiFi. Such databases include those used for customer relationship management, enterprise resource planning, personal e-mail, and digital libraries.

A key benefit of this model is support for arbitrary types of information and arbitrary data management systems. The server need only provide methods for querying or accessing the stored information over a network. Devices need means for retrieving and displaying items in standard formats, such as Web pages and JPEG images. Hence, this model is sometimes called *thin-client access*. Moreover, since the data is centrally maintained, access controls governing who is allowed to read and write various information items can be readily enforced. Data consistency is not an issue since all updates are performed directly at the server; devices that fetch data directly from the server always get the most recently written version.

FIGURE 2.1: Web access from a cell phone using WAP.

The main drawback of this approach is that data is inaccessible if a network connection to the server cannot be established or if the server is temporarily unavailable. Also, access time to the data is limited by the round-trip communication latency between the mobile device and the storage server, which can be substantial over certain wireless networks, such as cellular and especially satellite networks. Furthermore, such communication consumes valuable battery life on the mobile device and may incur network charges. For these reasons, the following alternative models explore different means for replicating information on mobile devices, where it can be accessed locally.

2.3   DEVICE–MASTER REPLICATION

Commonly, portable devices store full or partial replicas of data collections whose authoritative copy resides on a *master* site. The master source may be a shared server, such as a mail server, or a private computer to which the portable device is at least occasionally connected. These days, laptops, PDAs, music players, and even cell phones have enough storage capacity to replicate significant amounts of data from various sources. Even if a device has continuous connectivity to the master, entirely replicating databases, such as a person's address book and calendar, guarantees instant access to frequently used information and allows local searching. For mobile devices that have only occasional connectivity to information sources, replication is essential. For instance, a person's iPod may download music from the home PC (or indirectly from the Internet) only when connected by a USB cable. Without the ability to store music locally, the iPod would be useless.

Two broad approaches have been taken to ensure that a mobile device has ready access to critical data obtained from the master. One option is for devices to cache recently accessed data in local storage as a simple extension to the remote data access model. A related option is for devices to maintain an actively managed, user-visible replica of the master's data.

By caching data, mobile devices can amortize the cost of retrieving that data over several read operations and can retain access to that data when the device becomes disconnected from the master storage server. A laptop, for example, may cache files that have been fetched from a file server along with the set of recently browsed Web pages as shown in Figure 2.2.

In a caching model, data generally is fetched into a device's cache on-demand. That is, when the laptop user tries to access a file or Web page, the laptop's cache is first consulted to see if the data is already available. If the desired data is not cached (or if the cached copy is determined to be out-of-date and the user desires the most recent data), then the device may contact the appropriate server to fetch the data; in this case, the fetched data is stored in the cache for future access.

Devices can control the size of their caches and shrink or grow the cache based on their available storage. Since the cache size is limited, items may need to be discarded to free up space according to some cache replacement policy, such as removing the items that have been used least recently. For well-connected devices, a small cache may be sufficient to hold their working set of

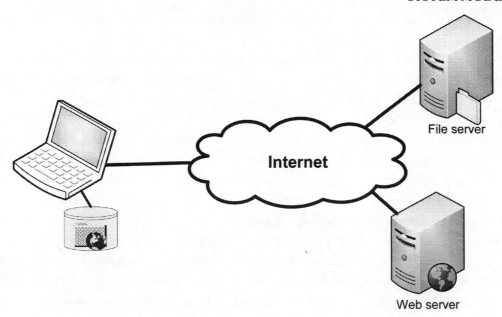

FIGURE 2.2: Laptop caching files and Web pages.

frequently accessed data and provide substantial performance benefits by avoiding much (but not all) communication with the server.

The drawback of on-demand caching is that information requested by a user will not be available if it is not cached and the server is not reachable. Such *cache misses* will occur for data that has not been recently accessed, or perhaps that has never been accessed. To minimize cache misses, *hoarding* (or *stashing*) can be used during periods of server connectivity to preemptively load data objects into a device's cache in anticipation of future use. *Hoard profiles* indicate which files or data objects a device is likely to access in the near future, perhaps while disconnected. Such profiles can be specified by users based on their anticipated needs or automatically generated from observations of past user behavior.

Replicating data on a mobile device is similar to caching in that the device stores data whose master copy lives elsewhere. However, the replication model differs from device-side caching in a number of key ways. One, a whole or partial data collection is copied onto a device at one time, rather than as individual objects are accessed, and explicitly refreshed periodically through a *synchronization* protocol. Two, attempts to read a data object fail if the data is not resident on the accessing device, rather than resulting in a cache miss and a remote access to the master. Three, data objects are implicitly added to a device's replica when new objects become part of the replicated data collection. Four, when a device deletes a replicated object, that object is removed from the data collection and all of its other replicas, rather than simply being discarded from the device's local storage.

Although a wide variety of replication techniques have been developed for nonmobile devices, including quorums and other techniques that provide strong mutual consistency guarantees, most of these are not applicable for mobile computing. To permit access to data replicated on disconnected devices, mobile systems rely on weaker consistency guarantees. In particular, users typically are permitted to read and write any data that is replicated on their devices without coordinating with other devices that may be sharing the same data. This *read-anywhere, write-anywhere* replication model is well suited to mobile devices with high-capacity storage but intermittent or weak connectivity and limited battery life. It is widely used for both consumer and enterprise applications [78].

Updates originating at a mobile device to a cached item are generally not only written to the cached copy but also written directly to the master server so that the updated item is immediately available to other devices. If a connection to the server is not currently possible, then updates may be performed locally and queued for later transmission. Updates made by other devices are not necessarily reflected immediately in a device's local cache. Although methods can be used for ensuring that caching clients always read the data that was most recently written, these techniques do not work for intermittently connected devices. Therefore, in mobile settings, a device usually is permitted to read old items from its cache, and thus the user may see stale information. Of course, if a user accesses data that only he updates, such as his personal calendar, then consistency is not an issue.

In the replication model, updates to replicated items are handled similarly. Data that is updated on a device are uploaded to the master site, and updates made on other devices are downloaded from the master site. Essentially, all updates are sent to the master site, which then distributes them to other devices holding replicas of the information. The process of communicating with the master to upload and download updated data objects is called *synchronization*. The term *reconciliation* or *reintegration* is also sometimes used for this process.

Synchronization takes place as connectivity permits and policy dictates. A device without wireless networking hardware, whose only means of communicating with a PC is through a wired *sync cradle*, for instance, synchronizes with the attached PC whenever the device is placed in the cradle (as shown in Figure 2.3). A device with wireless connectivity to the master site, such as a cell phone that synchronizes e-mail with a mail server, may synchronize its data periodically, say, every 5 minutes or when explicitly requested by a user.

One consequence of a write-anywhere replication model is that two users may independently update the same data item on different devices, thereby introducing *conflicting* updates. Even concurrent updates to different objects may conflict if, taken together, they violate some invariant that should hold on the data. In a device–master replication model, the master is responsible for detecting when two devices produce conflicting updates. In some cases, the master may be able to automatically resolve conflicts that arise, whereas in other cases, such conflicts may require human attention.

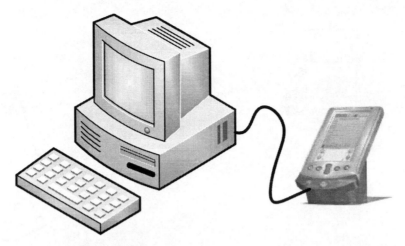

FIGURE 2.3: Synchronization between a PDA and home PC using a sync cradle.

2.4   PEER-TO-PEER REPLICATION

In a peer-to-peer replicated system, all devices holding a replica play (nearly) identical roles. In other words, there is no master replica. As with device–master replication, updates are generally propagated via pairwise synchronization operations. By relying only on communication between pairs of devices, peer-to-peer replication can effectively deal with varying connectivity between peers. Devices form an overlay network of arbitrary topology in which neighbors periodically synchronize with each other to propagate updates. Each node in the overlay network is a fixed or mobile device, and each edge represents a *synchronization partnership* between two devices with at least occasional network connectivity (as illustrated in Figure 2.4). Updated data objects flow between devices via the overlay network. Compared with the device–master model, peer-to-peer replication over arbitrary overlay topologies requires more complicated synchronization protocols but offers a number of key advantages.

With peer-to-peer replication, a device that belongs to a community of replicas can invite others to join the community simply by establishing local synchronization partnerships. The overlay topology can grow organically without informing other devices. Users need not even be aware of the full set of devices that are sharing data. Synchronization partnerships can come and go as long as the overlay network of replicas remains well-connected. If a mobile device opportunistically encounters another device that has data in common, these two devices can synchronize with each other without any prior arrangement or synchronization history.

The peer-to-peer replication process is tolerant of failed devices and network outages. If the master is temporarily unavailable in the device–master model, then devices cannot propagate new

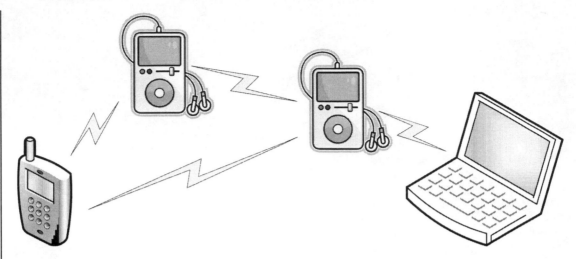

FIGURE 2.4: Peer-to-peer synchronization between mobile devices.

updates among themselves until the master recovers or reconnects. In a more general topology supported by peer-to-peer replication, such as a fully connected clique in which any device can directly exchange updates with any other device, the loss of a single device does not prevent updates from propagating along different paths.

One of the main advantages given for peer-to-peer replication is that it allows updates to propagate among devices that have internal connectivity but no connection to the Internet at large. For example, suppose that colleagues are holding an off-site meeting at a remote location without an Internet connection but want to collaboratively edit a document and share their edits between their laptops. The laptops may be connected by a local WiFi network or use point-to-point Bluetooth or infrared connectivity to exchange new versions of the document. As another example, teenagers may wish to send songs, ring tunes, and playlists directly between their portable music players or cell phones that are in close proximity.

Even when mobile devices are well-connected, nontechnical (e.g., political) concerns may lead organizations to favor configurations that do not rely on a master replica. Specifically, using peer-to-peer replication, also known as *multimaster replication*, puts all participants on an even footing. Studies of the role of information technology in disaster situations, for example, have shown that the various relief organizations that need to share emergency information wish to be viewed as equal partners. The peer-to-peer replication model supports collaborators operating as peers when managing shared data.

The principal cost of peer-to-peer replication is that it requires more complex protocols for ensuring that updated data objects reach each replica while efficiently using bandwidth. Also, update conflicts may be more prevalent than in the device–master model, and conflicts may be detected during synchronization between devices that did not introduce the conflicting updates. Overall, mobile users must deal with a more complex model resulting from the absence of a master replica, the lack of knowledge about the full replication topology, and decentralized conflict handling.

2.5   PUBLISH–SUBSCRIBE SYSTEMS

Publish–subscribe (or "pub–sub") systems are characterized primarily by their pattern of information dissemination. Small snippets of information, such as news articles, weather reports, and event notifications, are broadcast from a central site, the *publisher*, to a number of *subscribers*. Generally, information is group into topic-based *channels*, allowing devices to subscribe to items of interest. The information may reach subscribers directly or via other subscribers, who may be organized in a tree topology with the publisher at the root. The publisher and subscribers may be either mobile or fixed devices with wireless or wired communication capabilities.

From the perspective of mobile data management, a common and increasingly important scenario is a fixed publisher broadcasting information via wireless networks to mobile devices. For example, users may receive sport scores on their cell phones. Cellular providers offer such information services to attract customers and provide additional revue streams. Figure 2.5 depicts another example in which users receive news, weather, and sports on their watches.

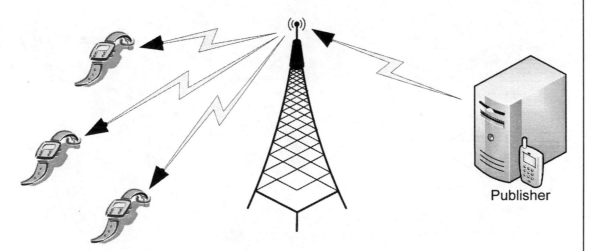

FIGURE 2.5: Smart watch receiving news, weather reports, and other notices that are broadcast from a central publishing site.

Like other replication models, once a user subscribes to a channel, such as news or weather, new items published to that channel are automatically replicated to the user's device(s). Such items are treated as read-only and created only at the publisher. Data replicated to mobile devices via a pub–sub system may be only of ephemeral interest; that is, the data is often discarded once they have been read by the user.

2.6   RELATED TECHNOLOGIES AND MODELS

A number of pervasive computing technologies and mobile systems that have received a fair amount of recent attention relate to the system models just presented. These are briefly discussed in the following subsections. These systems generally extend or combine the basic replication models.

2.6.1   Ad Hoc Wireless Sensors Networks

Much attention has been focused recently on small nodes equipped with computing resources, short-range wireless communication, and sensors for gathering temperature readings, motion, lighting, location, and a variety of other environmental properties. Such nodes can be configured, or automatically configure themselves, into ad hoc *sensor networks* that enable radically new monitoring and control applications. Sensors are being used for everything from monitoring volcanic/earthquake activity to tracking wildlife. In a sensor network, nodes may have short-range wireless

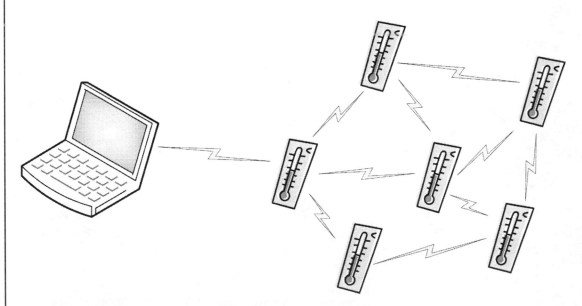

FIGURE 2.6.: Sensor nodes exchanging data via a mesh network and connected to a controlling PC.

connections over which they replicate sensor data and no or intermittent connectivity to a larger network [6]. As in the peer-to-peer replication model, sensor data is routed, and hence replicated, between nodes over a mesh of interconnected neighboring nodes (as shown in Figure 2.6).

Data management issues include replicating sensor data to the nodes that need to process it and aggregating such data as it propagates throughout the network. Generally, data is being collected continually and may demand real-time processing. Energy-efficient algorithms and protocols are required because sensor nodes often have severely limited battery life. In most cases, sensor nodes are homogeneous in form and function and hence operate as peers. However, the sensor network may include special nodes with increased computing, storage, battery, and communication capacity; such nodes may play a greater role in managing data collected from other sensor nodes. For example, ZebraNet sensors attached to wildlife regularly replicate data among nearby animals but only occasionally come within range of a long-range communication base station [55]. Thus, managing data in sensors networks involves a mixture of peer-to-peer and device–master models as well as in-network data filtering and aggregation.

2.6.2 Delay-Tolerant Networking

In a system in which two mobile devices that wish to communicate never (or rarely) are directly connected, messages can be routed between the devices via intermediate nodes over a *delay-tolerant network* [18, 39]. For example, Figure 2.7 shows a bus that picks up information at one bus stop and delivers it to a person waiting at a different bus stop. In this scenario, the two mobile devices that are communicating have only intermittent WiFi connectivity to the moving bus. Similar scenarios arise

FIGURE 2.7: Mobile vehicle serving as a data mule for devices at a bus stop.

in battlefield situations where the communication infrastructure connecting troops and vehicles may be changing and intermittent. Delay-tolerant networks can even be formed among humans through their chance encounters; these have been called *proximity social networks*.

Delay-tolerant networks resemble the overlay networks used in peer-to-peer replication. However, the objects being replicated between devices are messages rather than data items. Of course, such messages may include data such as Web pages, file contents, or news stories, but they are explicitly addressed to a particular device or set of devices. The topology of a delay-tolerant network may be fluid to take advantage of opportunistic encounters between devices. Unlike in the peer-to-peer model where all participating devices have a long-term interest in the data being replicated, the intermediate devices in a delay-tolerant networks are simply acting as good citizens in temporarily storing messages that are in transit to their destinations. Typically, once a message has been passed to a device that is presumed to be "closer" to the message's destination, the relaying device no longer stores the message. Because connectivity between devices cannot always be predicted, messages may be replicated among nodes in the network to increase their chance of successful delivery.

Delay-tolerant networks constructed from mobile devices face the challenge of deciding how best to route messages. In particular, two devices that come into contact must determine which of them is closer to a message's destination. Aggressively flooding messages among devices can be effective in maximizing the probability of delivery and minimizing the delivery latency. However, flooding also wastes bandwidth and consumes battery, both of which can be scarce commodities in mobile environments. It also presents another problem not faced in peer-to-peer replication, namely how to detect that a message has been delivered so that it can be safely discarded.

2.6.3   Infostations

A variant of the pub–sub model that incorporates ideas from delay-tolerant networking uses fixed devices, called *infostations* or *waystations*, as points from which mobile devices can receive information wirelessly [20]. Each infostation has both a wired connection to the Internet from which it receives information and a wireless network, typically WiFi, over which it broadcasts information to nearby devices. Infostations may be deployed in public places such as coffee shops and train depots.

Infostations provide convenient places where mobile devices can pick up (and potentially drop off) shared, replicated, contextually relevant information. Publishers in a pub–sub system might push selected channels to infostations from where the published information can be retrieved by subscribers that are passing by. For example, commuters might have news and weather information automatically downloaded into their laptops as they pick up their morning coffee (as shown in

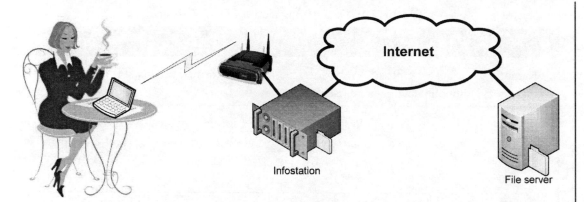

FIGURE 2.8: Internet-connected infostations deliver news that is picked up by a customer in a coffee shop.

Figure 2.8). Such information can even be tailored to the location of the establishment containing the infostation. Messages directed to particular devices, as in a delay-tolerant network, can also be replicated to infostations in places that are frequented by the target device. In this case, infostations serve as last-hop relays in the delay-tolerant network.

Because infostations are nonmobile devices, they avoid the storage and energy limitations faced by mobile devices. However, system designers do need to decide what information should be replicated to which infostations and for how long based on expected user needs and mobility patterns. Also, infostations should avoid sending information to mobile devices that have already received that information. Thus, they require efficient delivery mechanisms that are similar to the protocols used in peer-to-peer replication.

2.7   REPLICATION REQUIREMENTS

Table 2.1 summarizes the previous system models and their variations. These system models share, update, and distribute data in different ways. Thus, they place varied requirements on the replication protocols needed to support such systems. Table 2.2 summarizes the needs of the four basic system models regarding the following six environment and replication properties:

Continuous connectivity: Does the system only operate when devices are well-connected?
Update anywhere: Can any device update data items that then must be propagated to other replicas?

TABLE 2.1: Data sharing models of mobile systems

	DEVICES	DATA	READS	UPDATES
Remote access	Server plus mobile devices	Web pages, files, databases, etc.	Performed at server	Performed at server
Device caching	Server plus mobile devices	Web pages, files, databases, etc.	Performed on local cache; performed at server for cache misses	Performed at server and (optionally) in local cache
Device–master replication	Master replica plus mobile devices	Calendars, e-mail, address book, files, music, etc.	Performed on local replica	Performed on local replica; sent to master when connected
Peer-to-peer replication	Any mobile or stationary device	Calendars, e-mail, address book, files, music, etc.	Performed on local replica	Performed on local replica; synchronized with others lazily
Pub–sub	Publisher plus mobile devices	Events, weather, news, sports, etc.	Performed at subscriber's local replica	Performed at publisher; disseminated to subscribers
Sensor network	Sensor nodes	Environment data, e.g., temperature	Performed at nodes that accumulate and aggregate data	Real-time data stream at each node; routed to others over ad hoc network
Delay-tolerant networking	Any mobile devices	Messages (including files, Web pages, etc.)	Performed by message recipient	Message created by sender; routed to destination through intermediaries
Infostations	Publishers, infostations, plus mobile devices	News, messages, advertisements, events, music, etc.	Performed at infostation or on local replica	Performed at publisher; relayed to infostations; picked up by nearby devices

TABLE 2.2: Replication requirements for basic data-oriented system models			
	REMOTE ACCESS	**DEVICE–MASTER**	**PEER-TO-PEER**	**PUB–SUB**
---	---	---	---	---
Continuous connectivity	√√			√
Update anywhere		√√	√√	
Consistency		√√	√√	√√
Topology independence			√√	
Conflict handling		√√	√√	
Partial replication		√	√	√√

Consistency: Does the system require mechanisms to enforce consistency guarantees, such as eventual consistency (as discussed in Chapter 3)?

Topology independence: Is the connectivity between devices that replicate data unconstrained, i.e., defined by an arbitrary graph?

Conflict handling: May devices perform conflicting updates that need to be detected and resolved (as discussed in Chapter 6)?

Partial replication: Do devices wish to replicate some portion of a data collection?

In Table 2.2, each row presents the answers to one of the questions posed above, and each column presents the results for a given system model. Two check marks appear in a table entry if the answer to the corresponding question is "yes" when asked about the corresponding system model. A single check mark indicates that the feature is not required but is desirable. Not surprisingly, support for partial replication is a requirement of almost any mobile system due to a mobile device's limited storage and communication capacity, whereas the peer-to-peer replication model places the most demands on a replication protocol.

CHAPTER 3

Data Consistency

The *consistency* provided by a replicated system is an indication of the extent to which users must be aware of the replication process and policies. Systems that provide *strong consistency* try to exhibit identical behavior to a nonreplicated system. This property is often referred to as *one-copy serializability*. In practice, it means that an application program, when performing a read operation, receives the data resulting from the most recent update operation(s). Update operations are performed at each device in some well-defined order, at least conceptually. Maintaining strong consistency requires substantial coordination among devices that replicate a data collection. Typically, all or most replicas must be updated atomically using multiple rounds of messages, such as a two-phase commit protocol.

Relaxed consistency or *optimistic* models have become popular for replicated systems because of their tolerance of network and replica failures and their ability to scale to large numbers of replicas. These characteristics are especially important in mobile environments. Rather than remaining mutually consistent at all times, replicas are allowed to temporarily diverge by accepting updates to local data items; such updates propagate lazily to other replicas. Read operations performed on a device's local replica may return data that does not reflect recent updates made by other devices. Thus, users and applications must be able to tolerate potentially stale information. Mobile systems generally strive for eventual consistency, guaranteeing that each replica eventually receives the latest update for each replicated item. However, other stronger (and weaker) consistency guarantees are possible as discussed in this section.

3.1 BEST EFFORT CONSISTENCY

Many replication protocols simply make a best effort to deliver updates to all replicas. In such a case, read operations performed at different replicas may return different answers, even if no updates have been performed recently and hence no replication is in progress. Even systems that strive for eventual consistency, often fail to achieve it. Replicas may not converge to a mutually consistent state for any of number of reasons.

For one, all updates may not make it to all replicas. This can occur if updates are sent over a mostly, but not totally, reliable communication channel. For example, the Grapevine system propagated updates between replicas using e-mail, which is commonly viewed as being a reliable transport mechanism but is actually not [88]. Gossip protocols, as discussed in Section 4.3.3, make only probabilistic guarantees about message delivery.

Despite reliable delivery, replicas will not converge if

- updates are performed differently at different replicas (i.e., the application of an update is not deterministic);
- updates are applied in different orders at different replicas and are not commutable;
- replicas have different conflict resolution policies (as discussed in Section 6.3.1);
- metadata, such as deletion tombstones, are discarded too early;
- replicas lose or corrupt data, such as when a replica is restored from an old backup; or
- the system is improperly configured, such as when the synchronization topology is not a well-connected graph.

Replicated systems may intentionally provide no real consistency guarantees or unintentionally provide inconsistency through poor design, invalid operating assumptions, or misconfiguration.

3.2 EVENTUAL CONSISTENCY

A system providing *eventual consistency* guarantees that replicas would eventually converge to a mutually consistent state, i.e., to identical contents, if update activity ceased. Naturally, ongoing updates may prevent replicas from ever reaching identical states, especially in a mobile system where communication delays between replicas can be large due to intermittent connectivity. Thus, a more pragmatic definition of eventual consistency is desired.

Practically, a mobile system provides eventual consistency if (1) each update operation is eventually received by each device, (2) noncommutative updates are performed in the same order at each replica, and (3) the outcome of applying a sequence of updates is the same at each replica. Replication protocols that meet these requirements, as well as some that do not, are discussed in Chapter 4.

Eventually consistent systems make no guarantees whatsoever about the freshness of data returned by a read operation. Readers are simply assured of receiving items that result from a valid update operation performed sometime in the past. So, a person might, for instance, update a phone number from her cell phone and then be presented with the old phone number when querying the address book on her laptop. Techniques have been developed for providing stronger read guarantees while still allowing eventual update propagation. Some of these are applicable to mobile systems, and are discussed in the next sections.

3.3   CAUSAL CONSISTENCY

In a system providing *causal consistency*, a user may read stale data but is at least guaranteed to observe states of replicas in which causally related update operations have been performed in the proper order. Specifically, suppose an update originates at some device that had already received and incorporated a number of other updates into its local replica. This new update is said to causally follow all of the previously received updates. A causally consistent replicated system ensures that if update U2 follows update U1, then a user is never allowed to observe a replica that has performed update U2 without also performing update U1. If two updates are performed concurrently, that is, without knowledge of each other, then they can be incorporated into different devices in different observable orders. Protocols that use update logs, as presented in Section 4.2.1, can readily maintain causal ordering guarantees.

3.4   SESSION CONSISTENCY

One potential problem faced by users who access data from multiple devices is they may observe data that fluctuates in its staleness. For example, as explained above, a user may update a phone number on her cell phone and then read the new phone number from her PDA but later read the old phone number from her laptop. *Session guarantees* have been devised to provide a user (or application) with a view of a replicated database that is consistent with respect to the set of read and update operations performed by that user while still allowing temporary divergence among replicas [94]. For example, the "read your writes" guarantee ensures that the user only reads from replicas that have already received previous writes issued by the user. This solves the just-mentioned problem of reading an old phone number. Similarly, "monotonic reads" ensures that the user observes increasingly up-to-date data, "writes follow reads" ensures that a write operation initiated by a user is ordered after any writes previously observed by this user on any devices, and "monotonic writes" ensures a causal ordering on writes issued by the same user.

Unlike causal consistency, which is a systemwide property, session guarantees are individually selectable by each user or application. Application designers can choose the set of session guarantees that they desire based on the semantics of the data that they manage and the expected access patterns.

Session guarantees can be easily implemented on mobile devices, provided some small state can be carried with the user as she switches between devices. More practically, this state can be embedded in applications that access data from mobile devices. However, systems providing session guarantees on top of an eventually consistent replication protocol may need to occasionally prevent access to some device's replica. That is, availability may be reduced to enforce the desired consistency properties, which could adversely affect mobile users. One practical option is for the system to simply inform the user (or application) when an operation violates a session guarantee but allow that operation to continue with weaker consistency.

3.5   BOUNDED INCONSISTENCY

In some cases, bounds can be placed on the timeliness or inaccuracy of items that are read from a device's local replica, providing *bounded inconsistency*. For example, an application may desire to read data that is no more than an hour old, in which case, the system would guarantee that any updates made more than an hour ago have been incorporated into the device's replica before allowing a local read operation. Similarly, a system may enforce bounds on numerical error or order error as in the TACT framework [105, 106]. This requires replicas to know about updates made elsewhere and generally relies on regular connectivity between replicas. Thus, techniques for ensuring bounded inconsistency may not be applicable to all mobile environments.

3.6   HYBRID CONSISTENCY

Consider a system model in which updates originate at a single site, such as with device-side caching, pub–sub systems, or even perhaps device–master replication. In this case, applications can be presented with a choice of strong or weak read consistency. For applications running on a mobile device that require strong consistency, read operations can be directed to the master (or publisher), assuming the device is wirelessly connected to the master. Read operations that can tolerate stale data should be directed to the device's local replica, which may or may not contain the most recent updates from the master. Thus, when connected, applications face a trade-off between consistency and latency. When disconnected, strong-consistency read operations will fail, in which case, the application can either halt or accept potentially inconsistent data.

CHAPTER 4

Replicated Data Protocols

The designer of a replication protocol must deal with the following issues and fundamental questions:

Consistency: What consistency guarantees are desired and how are they provided?

Update format: Do replicas exchange data items or update operations?

Change tracking: How do devices record updates that need to be propagated to other devices?

Metadata: What metadata is stored and communicated about replicated items?

Sync state: What state is maintained at a device for each synchronization partnership?

Change enumeration: How do devices determine which updates still need to be sent to which other devices?

Communication: What transport protocols are used for sending updates between devices?

Ordering: How do devices decide on the order in which received updates should be applied?

Filtering: How are the contents of a partial replica specified and managed?

Conflicts: How are conflicting updates detected and resolved?

This chapter explores the range of practical answers to each of these questions (except for the last one on conflict handling, which is covered in Chapter 6). Consistency guarantees were discussed in the previous chapter. Unless otherwise stated, it is assumed in the following discussion that updates can be performed at any device (an update-anywhere model) and that eventual consistency is desired. That is, an application-initiated update operation is performed at a single device, causing that device to modify its local replica or possibly one or more remote replicas. The update is then propagated to other replicas over a wired network, wireless broadcast channels or device-to-device connections, which may be intermittent. A variety of replication protocols can be used for reliably and efficiently spreading updates.

4.1 REPRESENTING UPDATES

One of the most basic choices faced by a replication protocol designer is what to send between replicas to bring them into a coherent state. Each device's data store, whether a file system or database, provides an application programming interface (API) that applications call to perform update operations. Each operation, in turn, modifies the contents (and metadata) of one or more data items, such as files or rows in a database. Replicas can either exchange the operations performed by an application or the updated items resulting from those operations.

4.1.1   Operation-Sending Protocols

In an *operation-sending* protocol, replicas record and send operations to each other. Each device independently performs each received operation on its local replica. To achieve eventual consistency, replicas must not only receive the same operations (or correctly transformed operations as discussed in Section 4.4.6) but must also execute operations in a deterministic manner. Unless operations are *idempotent*, meaning that multiple executions have the same effect on the underlying data store, the replication protocol must ensure that operations are delivered exactly once to each replica. Also, the order in which operations are performed is often important (and suitable ordering techniques are discussed in Section 4.4).

4.1.2   Item-Sending Protocols

Alternatively, replicas can propagate data items, resulting in an *item-sending* protocol. Received updated items are added to a device's replica, normally replacing older items with the same unique identifier. The same item can be received multiple times by a device since replacing an item is an idempotent operation, but duplicate delivery should be avoided when possible since it wastes resources, which are precious on a mobile device. Ordering mechanisms (see Section 4.4) may be needed to determine if a received item is obsolete, i.e., is an older version of a previously received item.

4.1.3   Comparisons

There can be a fine line between operation-sending and item-sending protocols. In particular, items sent during replication can be viewed as very simple operations that completely replace the contents of a single item. However, operations can, and usually do, have higher-level semantics and may affect multiple items. More importantly, item-sending and operation-sending replication schemes differ fundamentally in what needs to be standardized among replicas. Whether standardizing on schemas or operations is preferable probably depends on the set of applications and devices that must be supported.

In item-sending protocols, replicas must share a common item layout, i.e., physical schema, since they exchange data items. However, different devices can run different applications with custom APIs for accessing different replicas of the shared collection. A pocket PC device, for instance, might run applications with more limited operations than the desktop PC with which it synchronizes. Moreover, APIs can evolve over time and new applications can be supported without affecting the replication process as long as the physical schema remains the same.

In operation-sending protocols, replicas must agree on the supported operations but are free to provide different implementations. Replicas may even have different physical schemas. For example, a pocket PC device may synchronize photos with a desktop PC but store them in a reduced resolution format. A photo editing application running on the desktop PC would send high-level operations, such as "crop this region of the photo," and the Pocket PC would need to apply such operations to its low-resolution version.

Most of the protocols discussed later can propagate items or operations. The generic term *update* is used to refer to either an updated item or an update operation.

4.2 RECORDING UPDATES

In the world of update-anywhere replication protocols, two basic schemes have been used to record the set of updates that need to be exchanged between replicas. In the *log-based* scheme, each device maintains a log of updates. Each update that originates at a device is stored in the device's log in addition to being applied to its local data store. Logged updates are delivered to other replicas as a background activity, and updates received from other devices by the replication protocol are added to the receiving device's log. In the *state-based* scheme, rather than keeping update logs, devices simply apply local update operations to their replicas. During replication, devices directly compare the contents of their replicas and exchange updated items.

4.2.1 Log-Based Systems

Systems that use update logs can either log operations or items, although such systems usually maintain operation logs. Thus, log-based systems are generally associated with operation-sending protocols. The replication process is concerned with propagating logged entries between devices, often ignoring the devices' data stores, as depicted in Figure 4.1. Different mechanisms for disseminating logged operations are presented in Section 4.3.

Replicas will converge to a consistent state provided that (1) each replica eventually receives and executes each update operation, (2) noncommutative operations are executed in the same order at each replica, and (3) operations have deterministic executions, i.e., an operation produces the same result at each replica given the same initial database state. This style of log-based replication has been used in a number of systems, including Coda for reintegration of disconnected clients (see Section 7.1) and Bayou as its sole means of ensuring replica convergence (see Section 7.3).

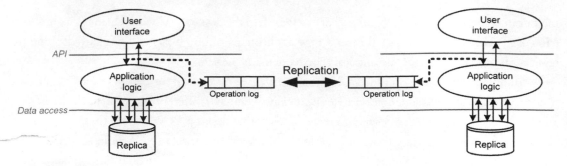

FIGURE 4.1: Log-based replication.

In general, the size of a device's operation log is unbounded and depends on the size of update operations, the update rate, and the propagation latency. Log-based systems must decide when logged operations can be safely discarded, i.e., when the log can be pruned. The answer is simple: a logged operation is no longer needed when it has fully propagated to all replicas or has been made obsolete by other operations occurring later in the log. Some systems provide log compaction mechanisms that discard obsolete operations, but determining obsolescence can be tricky for arbitrary operations. In a client–server system, such as Coda, clients can discard their logs once they successfully replay them to the servers. In a peer-to-peer system, such as Bayou, the mechanism for determining whether an operation is globally known is nontrivial, representing an additional implementation burden and source of bugs.

4.2.2 State-Based Systems

State-based systems invariably rely on item-sending protocols. Specifically, devices directly compare and exchange items from their local data stores, thereby avoiding the need for maintaining operation logs. This style of replication, shown in Figure 4.2, has been used in replicated file systems, such as Ficus, and even some database management systems.

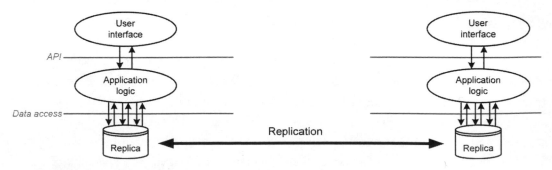

FIGURE 4.2: State-based replication.

To determine which items in a device's replica have been updated recently and also to correctly determine the latest version of an item during synchronization, the device's data store attaches metadata to each item. This additional metadata could be as small as a single modified bit (see Section 4.3.5), an update timestamp, or a version number. As with log-based replication, newly updated data items can be disseminated among replicas using a variety of mechanisms, including reliable multicast or peer-to-peer synchronization.

One issue that arises in a state-based replication protocol is how to handle deleted items. In a log-based system, delete operations are simply added to a device's log and propagate like all other logged operations. In a state-based system, if a delete operation simply removes an item from the initiating device's replica, this can cause problems. In particular, during synchronization between two replicas, if one replica holds an item and the other replica does not, the replicas cannot determine whether (1) the second replica never learned of the item and hence should receive the item or (2) the second replica deleted the item, in which case the first replica should also delete the item. This is known as the *create–delete ambiguity*.

The solution adopted by most systems is to mark items with a special "deleted" bit rather discarding them. Such deleted items are known as *tombstones* or *death certificates*. Tombstones replicate among devices just like other items. However, they are not visible to applications that access a device's replica.

4.2.3 Comparisons

State-based, item-sending replication protocols and log-based, operation-sending replication protocols can be compared along several dimensions. Table 4.1 presents a side-by-side technical comparison.

TABLE 4.1: Comparison of state-based and log-based replication

	STATE-BASED	LOG-BASED
Updates sent as	Data items	Operations
Metadata	Per-item modified bit, timestamps, or versions plus deleted bit	Operation log
Deleted items	Maintained as tombstones	Removed from database immediately, but delete operation logged
Physical item layout	Identical at all replicas	May vary across replicas
Operations (API)	May vary across replicas	Identical at all replicas

Although both styles of replication can provide the same high availability and eventual consistency, they differ in a number of key functional aspects.

The amount of metadata maintained in a state-based replication scheme is likely less since an operation log is not needed. For log-based systems, an operation log could easily grow larger than the replica itself. On the other hand, if delete operations are common, then the space needed to retain tombstones in a state-based scheme could be significant.

State-based systems can potentially reduce the overall message traffic. Whereas log-based systems always send each update operation to each replica, assuming log compaction is not possible, state-based replication naturally collapses a series of updates to each item into its latest version that is sent during synchronization. Thus, if a user is editing a document and saving it periodically, say, once per minute, but synchronizing with other users once per hour, the message traffic could be reduced by a factor of 60 compared with a log-based system that sends a steady stream of updates. On the other hand, log-based systems can often significantly reduce the size of their logs through effective log compaction techniques. Moreover, state-based systems typically send the entire contents of an updated item, even if a small portion of the item was modified. Sending high-level operations could be more efficient.

In essence, state-based replication is a special case of log-based replication in which the operations exchanged during replication are limited to single-item update and delete statements. This restriction allows the system to operate without an operation log, simplifies the implementation, and may provide performance benefits.

The downside is that state-based replication systems cannot provide the rich semantics or flexibility of log-based systems. Specifically, state-based systems lack three types of information that are captured in operation logs:

1. the sequence of operations that produced an item's current value,
2. the order in which items were updated, and
3. the set of items updated in a single operation (or transaction).

Applications that access data items may not care about this information, but it can be valuable in detecting semantic conflicts, enforcing atomicity, and preserving integrity constraints. For instance, log-based replication can, in theory, take into account the semantics of operations when detecting update conflicts, a potential advantage over state-based replication. However, as discussed in Section 6.2.7, exploiting operational semantics when detecting conflicts is difficult in practice. Additionally, a state-based replication protocol cannot guarantee that a set of updates that were performed atomically on the original replica will be performed atomically on other replicas.

4.3   SENDING UPDATES

This section considers a variety of protocols for exchanging operations or updated items, simply called *updates*, between replicas. For the most part, it assumes that replicas are interested in all items, that is, data collections are fully replicated. This assumption is relaxed in Chapter 5 where techniques for partial replication are discussed.

4.3.1   Direct Broadcast

Perhaps the simplest technique for disseminating updates is for a device that performs a local update operation to immediately send that update to all other replicas. For example, if all subscribers in a pub–sub system on are the same wireless network, the publisher can broadcast updates to each device. Multicasting updates to mobile devices in a cellular network is possible but expensive and not totally reliable [1]. In a mesh network, such as an ad hoc sensor network, devices could flood updates throughout the network by sending each received update to each neighbor. In a more general network setting, an updating device could send TCP/IP packets directly to each replica.

Directly sending updates between replicas avoids the need to log such updates. However, it often requires a device to know the complete set of replicas for a data collection. Moreover, devices that are currently unavailable, perhaps because they are out of network range or have a dead battery, may miss updates. To provide stronger delivery guarantees, more robust replication protocols are needed or the direct broadcast method must be augmented with one of the other protocols mentioned later.

For example, clients of the Coda file system, when in connected mode, multicast each update operation to the set of available servers that replicate the file being updated. If a server misses an update, this is detected by some client at a later time and a server-to-server repair mechanism is invoked.

As another example, Groove, a commercial collaboration system, directly sends update operations to each replica of a shared work space. Groove uses "relay servers" to support devices that are off-line and hence cannot currently receive updates and to accommodate devices that cannot directly communicate with each other, e.g., because they are behind firewalls. When a Groove client that is broadcasting an update determines that another replica of the shared space is unreachable, it sends the update message to that replica's relay server. The relay server simply queues each message it receives until the appropriate replica contacts it to retrieve the queued updates. If a relay server is unavailable, then a Groove client has no way of getting the update messages that are queued on that server even if other replicas of the shared work space are reachable.

4.3.2   Full Replica or Log Exchange

A simple and very robust protocol can be obtained by having pairs of devices periodically exchange the full contents of their replicas (in a state-based system) or logs (in a log-based system). When a device receives another device's replica (or log), it discards any duplicate items (or operations). For each of the remaining items, the receiving device needs to decide which version is the most recent; techniques for ordering versions of items are discussed later in Section 4.4. Obsolete versions are discarded. The device is left with a replica (or log) in which the data stored by it and its synchronization partner have been merged. Therefore, the two synchronizing devices become mutually consistent. Eventual consistency is easily achieved as long as each device holding a replica directly or indirectly communicates with all other replicas. Unfortunately, this simple protocol is not appropriate for most mobile systems since repeated exchanges of the same data items waste substantial bandwidth, consume CPU resources, and reduce battery life. Mobile devices require protocols that are both robust and network-efficient, i.e., that incrementally disseminate updated items.

4.3.3   Gossip Protocols

Randomized gossiping is one example of a replication technique that sends incremental updates between replicas. A variety of gossip protocols have been devised [12]. The basic characteristic of such protocols is that a device periodically sends a randomly selected update (or set of updates) to some other device. In one example, called *rumor mongering*, updates that are not widely known are treated as "hot rumors" and are actively propagated between devices. Devices randomly select updates from their list of hot rumors to send to other devices. These hot rumors can be stored either in the device's replica (in a state-based system) or in a separate log. When a device tries to deliver a hot rumor, it may discover that the updated item is already known to the target device. After a certain number of such failed delivery attempts, the device removes the item from the hot rumor list. Alternatively, updates can be removed from the hot rumor list after a fixed expiration period or based on some other policy. Such a gossip protocol is also call an *epidemic algorithm* because updates propagate in a style resembling epidemic diseases. Think of devices with hot rumors as being infected and actively trying to infect other devices (in a beneficial way).

Gossip protocols are simple to implement, make minimal assumptions about device connectivity, require only a small amount of metadata (the hot rumor list), and are reasonably robust. They support opportunistic encounters between devices. That is, two devices that have never previously been connected can exchange selected hot rumors while knowing almost nothing about each other.

The downside of gossip-style protocols is that an update may not reach all replicas. For instance, in the rumor mongering protocol described above, a rumor might propagate between a set of well-connected replicas and be declared cold before a disconnected mobile device has a chance to learn of it. To guard against this, devices may choose to occasionally send cold rumors. In any case, the protocol can make only probabilistic delivery guarantees and hence cannot ensure eventual consistency. In fact, as shown by Demers et al. [12], gossip protocols face a tradeoff between consumed bandwidth and residual delivery probabilities. Protocols that send updates more often or that less aggressively prune the hot rumor list can reduce the probability of incomplete delivery to a negligible amount but at the cost of sending more data.

Replication protocols that guarantee eventual consistency while maintaining tight bounds on the network resources required per update may be more suitable for intermittently and weakly connected mobile devices and are the focus of the remainder of this chapter. In some cases, other replication protocols can be used effectively in conjunction with a gossip-style protocol. In particular, devices might use a gossip protocol to quickly propagate recent updates among nearby neighbors and then a more expensive protocol, such as full replica exchange or anti-entropy, to ensure eventual delivery to all devices.

4.3.4   Message Queue Protocols

A reliable messaging system, such as IBM's MQ Series or Microsoft's SQL Service Broker, can also serve as a simple mechanism for propagating updates. The replica that performs an update operation simply places this update on queues to be delivered to all other replicas. In effect, the delivery queues managed by the messaging system serve as destination-specific operation logs that get pruned as updates are successful received.

Each replica need not have a direct connection to each other replica. A practical approach is to set up a multicast tree in which a replica sends updates to its child replicas, who, in turn, send each update to their children, and so on. Although it would be possible to have multiple multicast trees rooted at different nodes, the simplest scheme is to have a single tree. All updates are sent to the root of this multicast tree, which initiates the downward propagation. As long as message delivery is guaranteed, each replica will eventually receive every update.

Reliable multicast relies on having a multicast tree in which each replica occurs at least once. Efficient propagation relies on having a multicast tree in which each replica occurs exactly once. The replication topology must be a well-formed tree to avoid loops and ensure that each update operation is delivered only once. Configuring this tree as devices create and remove replicas can be challenging. Moreover, if the path from the root of the tree to some device involves intermediate mobile devices that are intermittently connected, then updates may propagate slowly.

Layering replication on top of a messaging infrastructure can potentially simplify the implementation but is also a cause for concern regarding eventual convergence. Even "reliable" multicast services are not entirely reliable. Years ago, the Grapevine system ran into this problem while using e-mail as its method for distributing updates [88]. Messaging services typically guarantee that either a message is delivered to the intended recipients or the sender is notified that it could not be delivered. The nondelivery case should be rare but must be handled by the replication layer nonetheless.

4.3.5 Modified Bit Protocol

Consider a situation in which two devices synchronize only with each other, such as a cell phone and home PC that share a copy of a person's address book. In this case, each device can keep track of which items have been updated by associating a *modified bit* with each data item. Initially, the modified bits for every item in the device's replica are set to zero. When an item is updated by a local operation, in addition to changing the item's contents, the item's modified bit is set to one. During synchronization, a device sends all of the items with a nonzero modified bit and receives such items from its partner device. When the synchronization protocol completes, all of the modified bits are reset to zero.

This modified-bit protocol is widely used in synchronization products for cell phones and PDAs, such as Palm's HotSync protocol. By maintaining only a single bit per item, it ensures that both replicas receive all updated items and provides eventual consistency. Additionally, if a given item is updated concurrently in two places, this condition can be detected by observing that the modified bits are set on both devices. However, as stated, this synchronization protocol only works for two devices.

The protocol can be extended for three or more replicas by having each device maintain a set of modified bits for each of its regular synchronization partners. A local update operation would then set the bits associated with each partner for each updated item. During synchronization, only the bits for the current synchronization partner are used to determine which items to send and are cleared at the successful completion of the process. If each device directly synchronizes with all other devices, at least occasionally, then this extended modified-bit protocol is efficient in that each update is sent exactly once to each device.

If two devices never directly communicate, that is, if two devices rely on intermediary devices for propagating updates between themselves, then a slightly adapted protocol is required. In particular, modified bits must be set not only for local update operations but also for updated items received during synchronization. This has the side effect of sending updated items over every path between two replicas, thereby wasting bandwidth but arguably increasing robustness and possibly reducing the replica convergence time.

In summary, the modified-bit protocol is a simple, low-overhead synchronization technique that works well for a small number of devices that directly communicate with each other or are configured in a star topology. It works less well for rapidly changing replica sets and does not allow incremental update exchange between devices that do not have an established synchronization partnership. In practice, storing a single modified bit for each item is much easier than managing an extensible set of modified bits for multiple synchronization partners, which explains why small mobile devices, such as cell phones and PDAs, are often limited to synchronizing with a single PC.

4.3.6   Device–Master Timestamp Protocols

The device–master model (as introduced in Section 2.3), compared with peer-to-peer models, allows a simpler replication protocol since fixed and mobile devices always receive updates from one source, the master. The modified-bit protocol is one example of a method for device–master synchronization but requires the master to maintain modified bits for each client device. This section considers alternatives that can handle any number of potentially changing clients.

One technique is for the master to assign *update timestamps* to each item that it updates locally or for which it receives an update from a client device. In other words, whenever the master receives or generates a new version of an item, it records the time obtained from the master's local clock as part of the item's metadata. The master's clock is assumed to be monotonically increasing, but need not be synchronized with the clocks on other devices or even be accurate with respect to the real time. So, for example, a simple *update counter* that is incremented and recorded for each new version could be used instead of timestamps.

Each client (or the master) records the time of its last synchronization with the master. When initiating a new synchronization session, the client first sends to the master this last synchronization time. The master then returns any items with more recent update timestamps, i.e., versions that have been produced since the last synchronization and hence are missing from the client. The master also returns its current clock value, which the client device records as its last synchronization time.

Having clients record last synchronization times frees the master from needing to maintain any long-term, client-specific synchronization state. Replicas can come and go without affecting the master's state. The master need only store per-item update timestamps (or update counters). Alternatively, the master could record last synchronization timestamps for each device holding a replica of the master's data collection. This would avoid the need for clients to provide this information at the start of each synchronization protocol and allow the master to initiate synchronization. A similar technique can be used for propagating updates that originate on a client device back to the master. In particular, a client can assign update timestamps to locally updated objects and also record the time of last synchronization with the master as measured by the client's local clock. During synchronization, the client sends any items updated since the last synchronization and then updates its last synchronization time. Note that unless clients and the master have synchronized clocks, which is unlikely in a mobile environment, each client must record two separate last synchronization times, one taken from the client's clock and one reported by the master.

4.3.7   Device–Master Log-Based Protocol

Another option for device–master replication is for client devices to maintain update logs. That is, each client stores its local update operations in a log and then sends the queued updates during synchronization. This requires no extra per-item metadata. After a successful synchronization with the master, a client can discard its complete log. So, log management is trivial. Coda, for instance, uses client logging during periods of disconnection and replays logged updates during reintegration.

For update operations that originate at the master or that are received by the master from a device and need to be propagated to other devices, master-maintained update logs are also possible, but less desirable. The master would either need to store a separate log for each client device or need to record per-client synchronization information to decide when updates can be discarded from a shared log.

4.3.8   Anti-Entropy Protocols

In a peer-to-peer replication model in which devices can obtain updated items directly from other devices without relying on a designated master site, additional challenges arise. Assume that each device can independently choose its synchronization partners and when to initiate or accept synchronization requests. Efficient pairwise synchronization between devices requires exchanging more metadata since a device cannot be certain about the state of its synchronization partners. Timestamp protocols, as discussed above for device–master replication, can be made to work correctly but are not bandwidth efficient. In particular, if a device records last synchronization times for each regular sync partner and uses this timestamp to decide which items to send, then each device will receive each update from each of its partners.

Ideally, a device should receive each update exactly once regardless of its communication patterns. This requires means for a device to determine which updated items may have been delivered to its sync partners via other devices. Additionally, a device should be able to synchronize with a new device that it encounters, assuming the two devices hold replicas of the same data collection(s), which also requires the ability to effectively determine what versions of items are already held by a replica.

A basic synchronization protocol that works for any pair of devices can be obtained by having one device send version metadata for all items in its replica so that the second device can determine what versions it is missing. Suppose that a laptop is synchronizing with a PDA over a wireless connection. The laptop would initiate the synchronization process by sending the unique identifier for each item that it stores along with the item's version and metadata, such as version vectors, that are needed to order versions (as discussed in Section 4.4) and detect conflicts (as discussed in Section 6.2). The PDA, upon receiving this information from the laptop, could then determine the items that it is missing and items for which the laptop has a more recent (or conflicting) version. The PDA would then explicitly fetch the missing versions from the laptop. The PDA can also determine which versions it stores that are unknown to the laptop and send these.

This *metadata exchange protocol*, also known as *anti-entropy* [12], makes minimal assumptions about the transport protocol used to send items during synchronization. Items can arrive in any order. If a message is lost, thereby causing an item to not arrive at the destination device, that item will be sent again during the next synchronization session. Thus, the protocol is robust enough to operate over lossy wireless networks. Importantly, it guarantees that a device receives an updated item at most once regardless of its synchronization partners and frequency. This synchronization protocol is particularly well-suited for state-based replication but could also be used for log-based systems by exchanging metadata about log entries.

Unfortunately, exchanging full metadata is expensive if collections are big, even if data compression is used over the network. Imagine sending information about all files in a large replicated file system during every synchronization session. Consider a system in which replicas synchronize with each other frequently so that updates propagate in a timely manner but items are infrequently updated. In this case, devices during synchronization will often already have consistent replicas and hence have no items to send. Sending the list of items and their metadata simply wastes networking and processing resources.

4.3.9   Anti-entropy With Checksums

One performance improvement on the basic anti-entropy protocol is for a device to first send a checksum (or one-way hash) computed over the contents of its local replica. The partner device could then compare the received checksum with its own computed checksum to see if the two replicas are identical. If the checksums match, then no further communication is needed at this time. If they do not match, then the full metadata exchange protocol can be used.

A variation on this scheme computes and exchanges checksums on sets of items, such as directories in a hierarchical file system [61]. This allows synchronizing devices to narrow down where differences in their replicas may reside and only send metadata for items in question.

Another variation is based on the observation that synchronizing devices, if not totally consistent, are generally missing but a few recent updates. Thus, devices can maintain a relatively short log of recent updates. During synchronization, devices first exchange metadata for the items in their logs, then send missing log entries, and finally, after adding the received items to their replicas, compute and exchange checksums. The checksums will fail to match only in the case that a device was missing an update that was not in the other device's log. In this rare case, resorting to the expensive anti-entropy protocol is acceptable.

Yet another variation uses a technique called *peel-back checksuming* [12]. Suppose that devices can deterministically order their items using timestamps or maintain an ordered update log. Two synchronizing devices first exchange and compare checksums. If the checksums do not match, then the device with the latest timestamped item (or latest log entry) sends that item to its partner. The receiving device, after applying the received item, incrementally recomputes the checksum for its replica. If the new checksum still does not match the other device's checksum, then this process of sending items in reverse timestamp order and comparing updated checksums continues until the devices have matching checksums. This technique reduces the network traffic but may increase the overall synchronization latency.

4.3.10 Knowledge-Driven Log-Based Protocols

Having each device maintain *knowledge* about the set of versions or operations that have been incorporated into its local replica enables a class of efficient, knowledge-driven replication protocols. Let us start by considering log-based systems in which each device maintains a log of every update operation that it has received. Conceptually, a device's knowledge is the set of unique identifiers for all operations that are stored in its local log. Whenever an update operation is performed on a device's replica and added to its log, that operation's ID is added to the device's knowledge. A device always knows about the updates that it has initiated and learns about other operations when they are received during synchronization.

By exchanging their knowledge, devices can ensure that they do not receive redundant updates, even in a system that permits peer-to-peer replication over an arbitrary overlay network. Specifically, one device, the "target," initiates synchronization by sending its knowledge to one of its synchronization partners, the "source." The source device can determine from the target's knowledge which operations in the source's log are not already known to the target, i.e., are not also included in the target's log. These operations are sent from the source to the target. Received operations are added to the target device's log and applied to its replica. After synchronization, the source need not remember the target's knowledge, and indeed, doing so would be fruitless since this information would become out-of-date as the target synchronizes with other partners. This is an example of a "pull" protocol since updates flow in one direction to the device that initiated the synchronization. To arrive at a mutually consistent state, two devices must pull from each other.

This *knowledge-driven protocol* resembles the metadata exchange protocol described in the previous section and can be considered another example of an anti-entropy protocol. The key difference, however, is that a device's knowledge can be represented in a much more compact form than the metadata for each item. In particular, the Bayou system demonstrated that knowledge can be condensed into a vector containing an entry for each replica rather than an entry for each item. In most systems, the number of replicas is substantially smaller than the number of items in a data collection, perhaps by several orders of magnitude.

A *knowledge vector* is a data structure containing a set of <replica, accept-stamp> pairs. Suppose that a device, when performing an operation on its local replica, assigns a unique *accept-stamp* to the logged operation consisting of the unique identifier of the device's replica along with a counter that gets incremented for each operation performed by the device (but not for each operation received during synchronization). Timestamps can be used instead of update counters as long as they are taken from a monotonically increasing clock, and no two updates from the same device are assigned the same timestamp. Each operation's accept-stamp is logged with the operation and transmitted during device-to-device synchronization. A device's knowledge vector contains a single entry for each replica (i.e., device) in the system. The knowledge vector entry associated with a replica indicates the accept-stamp of the latest update operation initiated at that replica that is known to the local device.

As long as update operations originating at a device are received by other devices in the order in which they were performed, a knowledge vector can precisely represent all of the operations that are stored in a device's log. For example, if Bill's laptop receives an operation whose accept-stamp indicates that it is the fifth update performed by Joe's PDA, then Bill's laptop must have already received the PDA's first four updates. Thus, after receiving this operation (either directly or indirectly from Joe's PDA), the knowledge vector for Bill's laptop should contain the entry <Joe's PDA, 5>. A partner receiving Bill's laptop's knowledge during synchronization can conclude that the laptop already knows about updates one through five from Joe's PDA.

Using knowledge vectors, the synchronization protocol operates as follows. The target device sends its knowledge vector to the source. The source runs through all of the entries in its log, starting at the beginning, until it finds an operation whose accept-stamp is unknown to the target. To determine whether an operation is known, the source uses the accept-stamp's replica ID as an index into the target's knowledge vector and then compares the counter of this knowledge entry with the counter in the accept-stamp. Note that known and unknown operations may be interleaved in the source device's log, and thus the source must continue to check accept-stamps against the target's knowledge. By sending operations in the order that they appear in the source's log using a transport that provides reliable, ordered delivery, such as a TCP or HTTP connection, the target can immediately append received operations to the end of its log and update its knowledge vector using the operations' accept-stamps. A device is free to reorder the operations in its log (as discussed in Section 4.4) as long as it maintains the property that two operations that originated at the same replica appear in the log in the order in which they were performed.

As with the metadata exchange protocol, this knowledge-based, peer-to-peer protocol is efficient, robust, and flexible. It guarantees that each operation is received by each device exactly once regardless of the synchronization partnerships between devices. If the synchronization process between two devices is interrupted, it can be resumed later. Any operations received before the interruption can be logged and applied to the receiving device's replica. As long as the synchronization topology graph is well-connected, it is guaranteed that each replica will eventually receive all updates. Loops and multiple paths between devices are allowed (and even encouraged). Therefore, each replica can unilaterally decide what other replicas to synchronize with and on what schedule.

The topology of which replicas synchronize with which other replicas is flexible and can vary over time. The full set of replicas need not be known to anyone. Adding a new replica can be achieved by simply creating a replica with an empty log and zero knowledge; this replica's device will receive a full set of items when it first synchronizes with another device.

For mobile devices, maintaining full operation logs is usually infeasible. One option to reduce logging requirements is for devices to run a distributed protocol to determine which operations have fully propagated to all devices. Devices can safely remove such operations from their logs. Distributed garbage collection protocols are often too complicated or too expensive for mobile environments.

Another option is to allow each device to independently discard operations from the beginning of its log. Provided that a prefix of the log is pruned, a knowledge vector, known as the *omitted vector*, can precisely characterize the set of discarded entries. Each device that has pruned its log must persistently store an omitted vector. For example, a device that completely erases its operation log would store its current knowledge vector as its omitted vector. During synchronization, the source device can compare its own omitted vector against the target device's knowledge vector to determine whether it has discarded any operations that are not yet known to the target. If so, then the source can either decline the target's synchronization request or the two devices can resort to a more expensive synchronization protocol such as a metadata exchange protocol (as in Section 4.3.8) or having the source device send its full replica (as in Section 4.3.2).

4.3.11 Knowledge-Driven State-Based Protocols

Some mobile devices, such as cell phones and PDAs, are not equipped to store or manage any sort of operation log. To accommodate such devices, the WinFS system demonstrated that a variation of this knowledge-driven protocol can be used for state-based systems in which devices do not maintain update logs [62]. Each item in a replica is stored with a unique version number that is updated whenever the item is updated. The version number is assigned by the device that performs an update operation and propagates with the updated item during synchronization. As discussed above, each device maintains knowledge in a compact representation, namely as a knowledge vector. However, for state-based systems, the knowledge vector is conceptually a set of versions rather than a set of operations. Each entry in a device's knowledge vector is a <replica, version number> pair indicating the highest version known to the device that originated at the given replica.

The knowledge-driven, state-based synchronization protocol is essentially the same as that described above, but with a few wrinkles. As usual, a replica first sends it knowledge to its synchronization partner in a request message. The partner replica then searches its replica (rather than a log) for versions not included in the received knowledge and sends these data items to the target device. Received items replace those items with the same unique identifier in the target's replica, assuming the target determines that a received version supersedes its own version (as discussed in Section 4.4). However, two key differences arise compared with a log-based scheme: (1) updated items may be received in any order and (2) overwritten versions of an item may never be received.

Because no log exists to maintain the causal order between versions, updated items can be sent by the source in any order. Typically, the source device runs a query to enumerate the items in its replica. To avoid having to sort the complete replica by version number, which may not be possible because of memory limitations on a small mobile device, items are sent to the target in whatever order is returned by the query processor. Thus, the target may receive an item assigned version 5 by Joe's PDA before it receives a different item that was also updated by Joe's PDA and assigned version 3. This means that when receiving Joe's PDA's version 5, the target device cannot update its knowledge vector to contain the entry <Joe's PDA, 5>. This would incorrectly imply that the device has received versions 1 through 5 from Joe's PDA.

To cope with this issue, WinFS introduced the notion of a *knowledge exception*, a version that is explicitly added to a device's knowledge that is not covered by the device's knowledge vector. In the previous example, when the target received version 5 assigned by Joe's PDA, it would add a knowledge exception for this version. That is, the target's knowledge vector might contain the entry <Joe's PDA, 2> plus an exception for <Joe's PDA, 5>. If the device later receives version 3 and version 4 from Joe's PDA, then it can update its knowledge vector to contain the entry <Joe's PDA, 5> and remove the knowledge exception. Note that when sending its knowledge during synchronization, a device includes both its knowledge vector and knowledge exceptions. This technique allows items to arrive in any order and thus can tolerate lost, reordered, or duplicated messages. The main cost of out-of-order delivery is that a device's knowledge increases in size because of knowledge exceptions, but such increases should be temporary.

A more serious problem stems from the fact that state-based synchronization protocols only send the latest version of an item (because that is all that is stored in a device's replica). Consider the case where Joe updates an item, such as an address book entry, on his PDA, producing version 4, and later updates the same item, producing version 5. Suppose that Bill's laptop has a knowledge vector indicating that it knows up to version 3 from Joe's PDA. When Bill's laptop synchronizes with Joe's PDA (or some other device that had already synchronized with Joe's PDA), it will receive version 5 of this updated item. Bill's laptop will therefore create a knowledge exception for this version. However, the laptop will never receive version 4, which was overwritten by version 5. Thus, Bill's laptop will be left with a knowledge exception indefinitely unless other steps are taken.

To cope with this issue, WinFS added a step to the end of the synchronization protocol in which the source device sends its current knowledge as *learned knowledge*. The key insight is that at the end of the synchronization process, assuming the target device successfully receives all of the items that were sent by the source, the target should know about all of the versions that were known to the source. Thus, the target device applies the received items to its replica and, if no problems are encountered or messages lost, adds the received learned knowledge to its local knowledge. Two knowledge vectors can be combined by taking the maximum version for each entry. With this application of learned knowledge, exceptions that are added to a device's knowledge as items are received during synchronization can be discarded when the synchronization process terminates successfully. Thus, devices end up with compact knowledge vectors in the steady state.

4.4 ORDERING UPDATES

To reach eventual consistency in a state-based replication protocol in which devices send updated items, devices must agree on which version of each item is the latest version. In other words, when receiving an item via the replication protocol, a device must decide whether the received version is later than its stored version of this item. If yes, the device should replace its version with the newly received version. If the received version is not more recent, then the device should ignore it. Unless devices make identical decisions on how to order updates, i.e., versions of items, they will not converge to a consistent state.

Similarly, in a log-based protocol, when two devices exchange log entries, newly received updates cannot simply be appended to the end of a device's log. The order in which noncommutative operations are applied to a replica may be significant. Since update operations are not always commutative, i.e., different execution orders can produce different end results, devices invariably need to agree on the order in which updates should appear in their merged logs and executed. Thus, techniques for deterministically ordering updates are an important component of most replication protocols.

4.4.1 Ordered Delivery

One approach is for the replication protocol to ensure that updated items (or update operations) reach all devices in the same order. In this case, each device can simply apply the updates as they are received. This may be practical in a pub–sub or other system where updates originate at a single device and propagate via a reliable, ordered transport protocol, but guaranteeing ordered delivery in other situations is not generally feasible. Ordered multicast protocols, for instance, require multiple rounds of messages and make assumptions that do not often apply in mobile environments.

4.4.2 Sequencers

Another approach is for one designated device, called a *sequencer*, to assign sequence numbers to items or operations. Each update operation causes the updated item to receive a new, higher sequence number (or update counter). When receiving updates during synchronization, devices can easily and deterministically order them according to their assigned sequence numbers. This scheme may work well for configurations in which all updates flow through a single device before being disseminated to other devices, such as in a hub-and-spoke synchronization topology or a device–master replication scheme. The downside, of course, is that the sequencer may become a bottleneck, and failure of the sequencer prevents updates from propagating. Alternatively, replicas could run a distributed agreement protocol, such as Lamport's Paxos protocol. However, this would require a replica to contact a majority of the other replicas, which is often not possible in mobile systems.

4.4.3   Update Timestamps

A commonly used decentralized technique that works with almost any replication protocol is to assign a timestamp to each update. When an item is updated, the updating device records the time as indicated on its local clock. This *update timestamp* stays with the item or operation as it propagates via the replication process.

Given two versions of an item in a state-based replication scheme, such as its stored version and one received during synchronization, a device can compare timestamps and keep the version with the latest timestamp. This assumes that each device's clock is monotonically increasing. It also assumes that two updates to the same item by different devices are not assigned the same timestamp, which is easy to guarantee by appending a device's identifier to timestamps that it generates.

Importantly, with this use of timestamps for ordering versions (or operations), no assumptions need to be made about the transport protocol, other than eventual delivery. Updates can arrive in any order and can be delivered multiple times. Clock synchronization is not required for eventual consistency, but reasonably accurate clocks are desired for other reasons. If a device has a clock that is too far in the past, its updates may be ignored, whereas a device with a clock too far in the future may prevent other devices from updating items that it has created.

Logical clocks can ensure that versions are ordered according to the "happens-before" relation as defined by Lamport [53]. For example, suppose that a file is updated on a person's laptop and then replicated to his home PC, where it is updated again. Ideally, the version produced on the PC should be retained as the latest version in all replicas. This requires the PC version to be assigned a larger timestamp, which is guaranteed if timestamps are taken from logical clocks that are updated using the algorithm suggested by Lamport.

4.4.4   Update Counters

Per-item *update counters* can also be used to produce an ordering on versions that is consistent with the order in which updates are performed. Stored (and sent) with each item is an integer that counts the number of updates that have been performed on the item. When an item is first created, its update counter is set to one. Subsequent updates by any device increment the item's update counter. A received item is applied to a device's replica, provided its associated update counter is greater than that of the replica's stored version. One minor complication is that two devices that concurrently update an item will produce versions with the same update counter. To consistently order such concurrent updates, the item's metadata should also include the identifier of the last updating device; these device identifiers can be used to distinguish different versions with identical update counters.

4.4.5 Version Vectors

Version vectors are another ordering technique that preserves the causal order of updates to an item. Conceptually, a *version vector* is a set of update counters, one per device. Each item is assigned its own version vector. When an item is created, its version vector contains a zero value for each device except for the creating device, for which it contains a one. To save space, entries with zero-valued counters can be omitted from a saved or communicated version vector. When a device updates an item, the device increments its own entry in the item's version vector. One version vector is said to *dominate* another if, for each device, the counter in the first version vector equals or exceeds the counter in the second version vector and, for at least one device, the first version vector's counter is greater than that of the second version vector. A received version of an item is ordered after the previously stored version of this item if the received version vector dominates the stored version vector for the item. Version vectors, although more complex than simple update counters, cannot only consistently order versions but also detect versions that were produced concurrently and hence are considered conflicting (as discussed in Section 6.2.4).

Strictly speaking, each entry in a version vector need not be an update counter. The entry for a device need only increase on each update performed by that device. So, variations on the basic version vector scheme are possible in which timestamps or other types of counters are used in place of update counters.

4.4.6 Operation Transformation

Another approach that has been extensively studied but rarely, if ever, used, involves *operation transformation* [17, 59, 79]. Rather than undoing operations that arrive "out-of-order," each replica performs operations in the order in which they are received, perhaps modifying an operation such that its effect on the replica's state is that same as if it had been received in the proper order.

For example, suppose two people are concurrently editing a replicated text file using a collaborative document editor. They both start with identical copies of the file. Alice inserts the six character adjective "mobile" at character position 10 in the file while Bob deletes a five-character word, e.g., "fixed," that starts at position 22. Alice immediately performs her insert operation on her local replica of the file and then later receives the delete operation from Bob. Suppose that all of the replicas agree that Bob's delete operation occurs after Alice's insert operation using any of the techniques previously discussed for producing a global update order. When Alice's replica receives Bob's delete operation, it cannot simply delete five characters starting at position 22 since that would result in deleting a different five characters than were deleted by Bob. Instead, Alice's replica, before executing the operation received from Bob, transforms this delete operation to account for the fact that Alice already performed a concurrent insert operation. Specifically, Bob's operation is transformed into an operation that deletes five characters starting at position 22 + 6 = 28. Similarly, when Bob receives Alice's insert operation, it must transform it into an alternative operation whose effect is equivalent to Bob's replica having performed the insert operation *before* its own delete. In this example, no transformation is needed by Bob's replica, although the operations were received and performed out-of-order. Bob can simply insert "mobile" at position 10 in the file, as did Alice. If Alice's insert had been at a different position in the file, say, after position 22, then Bob would have needed to transform this operation appropriately. Alice and Bob will both end up with identical replicas of the shared file despite performing different operations in different orders.

Operation transformation requires the same techniques for logging and ordering operations as other log-based replication protocols, but has the benefit of allowing replicas to perform operations in any order. On the downside, application designers are burdened with the task of producing appropriate transformations for all possible operations and their orderings. Such transformations must take into account not only the semantics of the operations but also their parameters and intended ordering. For n operations, n^2 operation transformations are required. As an application evolves to support additional functionality, new operations and associated transformations will likely be needed. As one might imagine, producing and maintaining correct operation transformations is a challenging endeavor. For this reason, transformations have only been devised for simple data structures with a small number of operations, such as text strings. Even so, most of the papers published on this subject include bugs in their reported transformations [38]. Moreover, it is not always possible to produce semantically correct transformations for all concurrent operations. In such cases, the concurrent operations will be treated as conflicting and must be resolved as discussed in Chapter 6.

4.4.7   Other Ordering Issues

One problem in log-based systems is how to deal with operations that are received before their global order is known. For instance, updates performed on a disconnected laptop can be logged locally, but their global order relative to other concurrently executed operations cannot be determined until the laptop is able to communicate with other replicas. Some reliable multicast systems wait to deliver messages until a global order can be established. This is not practical in a replicated data system since it would mean that locally performed updates are not immediately visible, even to the local replica. Users would find it very disturbing, for example, to update their calendar on a disconnected laptop and have those updated entries not show up until the laptop reconnects.

Thus, systems using log-based replication generally apply local updates immediately to the local replica. This is referred to as "tentative" execution since the local updates may need to be undone and performed in a different order as new update operations are received from other replicas and the global order is agreed upon. The system provides mechanisms not only for determining a global order on operations but also for undoing operations that were tentatively executed out of order. This is a substantial implementation effort that is unnecessary in state-based replication systems where whole contents of data items are exchanged during replication and the order in which different items are updated at the target replica does not matter.

If the system understands which operations are *commutative*, then it can avoid undoing/ redoing operations whose order of execution produces the same result as the global order. The Ice-Cube system, for instance, provides an API for indicating ordering constraints on operations [43]. Each replica is then free to choose a local execution schedule that is equivalent to that defined by the global order, that is, a schedule that does not violate any of the specified constraints.

* * * *

CHAPTER 5

Partial Replication

As shown in Section 2.7, many mobile scenarios involve devices that store *partial replicas* containing only a subset of the items in a data collection. Partial (or filtered) replication supports devices with limited storage and can reduce communication by selectively disseminating items only to the devices where they are needed. The replication protocols and techniques described above ignore issues that arise with partial replicas. These issues include

- How do devices decide which items to retain in their partial replicas, i.e., specify the items of interest?
- When and where are filters applied to items propagating between replicas?
- What happens when a device changes its interests?
- What happens when items are updated causing them to move in or out of a device's interest set?
- What constraints must be placed on synchronization topologies that include partial replicas?

This chapter addresses these issues for a variety of partial replication schemes characterized primarily by the means in which devices select the items of interest.

5.1 ACCESS-BASED CACHING

In a device-side caching scheme (as explained in Section 2.3), the number of items replicated onto a device is limited by the size of its cache, and the choice of which items to cache is generally determined by the device's access patterns. For example, a cache in one's laptop or PDA may contain recently accessed Web pages. The main issue is how to ensure that cached items remain (reasonably) up-to-date. When an item is updated elsewhere, a device should eventually learn that its cached copy is invalid or else receive the updated contents. Similarly, when a cached item is updated locally, the update should eventually propagate to other devices that cache or replicate this item. This is a replication problem, and many of the previously discussed replication protocols could be used to provide eventual cache consistency.

One commonly used technique involves a server machine (or designated full replica) recording which items are cached at other devices (i.e., partial replicas). When the server receives an updated item (by way of any replication mechanism), the server immediately informs devices that have the item cached using what is known as a *callback*. Similarly, when a device performs a local update operation, it can immediately send the updated item to the server (or group of devices). This is an example of the direct broadcast scheme discussed in Section 4.3.1. It propagates items as quickly as possible, thereby minimizing cache inconsistency.

One concern, however, is that a device that is down or disconnected will miss callbacks. To recover, some systems take the conservative approach of discarding all cached entries when a device is rebooted or reconnected, which may be a common occurrence for mobile devices. A more attractive alternative is to use a replication protocol that can tolerate intermittent connectivity and deal with dynamic cached contents, such as the modified-bit protocol (Section 4.3.5), the update timestamp protocol (Section 4.3.6), a metadata exchange protocol (Section 4.3.8), or a knowledge-driven protocol (Section 4.3.11). The replication protocol could periodically be initiated by a device, in addition or instead of relying on server-provided callbacks. Moreover, peer-to-peer replication could allow caching devices to receive updates (or invalidations) from a number of other devices.

Another problem with callback schemes and most replication protocols is that the server must be aware of the items that are cached on each device. During synchronization with a partial replica on a mobile device, the server wants to avoid sending updated items that the device no longer stores (or perhaps never stored). This is problematic for two reasons. First, mobile devices that discard cached items to free up storage must inform the server, thereby requiring extra communication. If the device discards cached items during periods of disconnection, the server's records may become out-of-date. Second, in peer-to-peer scenarios, devices may obtain items from multiple sources and thus may require additional communication to establish callbacks with the designated server.

Replication protocols that do not rely on callbacks circumvent these problems of maintaining server state about device caches, thereby allowing devices to operate while disconnected and unilaterally modify their cached contents. However, they still face the fundamental problem that the set of items cached on a device cannot be compactly conveyed. Consider knowledge-driven protocols (Section 4.3.11). A device could conceivably retain a knowledge vector characterizing the versions of items in its local cache and then pass this knowledge to a synchronization partner to poll for items that have been updated. However, the partner device would need additional information about the set of cached items to determine which updates/invalidations to return, that is, to avoid returning information about noncached items. For network efficiency, knowledge-driven protocols assume that the synchronizing devices store the same sets of items (or at least are aware of each others' interests), which is not a valid assumption for access-based caches.

Anti-entropy protocols in which a device sends metadata about all items in its partial replica (called metadata exchange protocols in Section 4.3.8) would appear more suitable for device-side caches. However, such protocols are expensive if the amount of metadata is large. To minimize communication, the device that stores the fewest items, which would be a caching device when synchronizing with a full replica, should initiate synchronization.

5.2   POLICY-BASED HOARDING

Hoarding is technique that preloads items into a device-side cache so that the items are available during periods of disconnection. While a device is connected to a server machine or peer replicas, the hoard process, sometimes called a *hoard walk*, decides which items should be stored in the device's cache and explicitly fetches (or replicates) such items. This decision may be based on system-defined or user-specified policies that can take into account the set of items needed to perform specific tasks or other user preferences. For instance, a person who is writing a paper at night while away from the office and hence disconnected from the company's network-based file server could specify in her profile that all of the files related to this paper should be retained on her laptop, including not only the main document but also the figures, related work, and experimental results that are being incorporated.

Before disconnecting, a device should synchronize with the server machine or one or more peer devices to ensure that its hoarded items are up-to-date. When a device reconnects after a period of disconnected operation, it should once again synchronize to propagate locally updated items and receive updated items. As with access-based caching, a variety of synchronization protocols can be used.

5.3   TOPIC-BASED CHANNELS

In a pub–sub system (Section 2.5), published information may be categorized by topic (or other criteria) into *channels*. Devices can subscribe to just those channels of interest. For example, one person may want to receive sports scores on their cell phone, whereas another only wants current events. A device receives all of the items published in any of its subscribed channels.

For purposes of replication, each channel can be treated as an independent data collection. While broadcast or tree-based multicast techniques are typically used in pub–sub systems, any replication protocol could be used, and arbitrary synchronization topologies can be supported. For example, gossip-style protocols (Section 4.3.3) could disseminate breaking news stories among mobile devices that encounter each other. In such a scenario, full propagation to all devices may not be necessary.

Using peer-to-peer synchronization, mobile devices can share channel data even when a connection to the publisher is unavailable or accommodate multiple independent publishers. A device could have different synchronization partners for different channels or one set of partners used for all of its subscribed channels. In the later case, two devices during synchronization will send items for channels in which they have common interests and ignore other channels. With a knowledge-driven protocol, a device should maintain a separate knowledge vector for each channel to which it subscribes. In this case, an article that is published in multiple channels will be delivered multiple times. Alternatively, the device could maintain a single knowledge vector that records items it has received over any subscribed channels. It could then obtain new items for multiple channels in one synchronization session. However, in this case, two problems arise. First, when a device synchronizes with a partner that contains some but not all of the same subscribed channels, the device cannot accept learned knowledge (as described in Section 4.3.10) since this may cause it to miss updates on other channels. Second, a device that changes its subscriptions may need to discard its knowledge and essentially restart.

5.4 HIERARCHICAL SUBCOLLECTIONS

For data collections that are organized hierarchically, devices may choose to replicate some subset of the collection, called a *subcollection*. As an example, a person may want all of his e-mail folders available on his desktop PC but only the messages in his inbox replicated into his cell phone. Similarly, a file server may store a person's complete file system, whereas a laptop stores only those files in specific directories. In these examples, the set of items partially replicated on the mobile device can be easily named (e.g., the name of an e-mail folder or file directory).

In a device–master model where the device holds a partial replica of data stored on the master, synchronization is straightforward. Modified bits, update timestamps, or other methods can be used to track which items have been updated on the device or master or both. When synchronizing to the device, the master only sends updated items that are included in the device's subcollection. If operation logging is used, the master must be able to determine which operations affect items in the subcollection.

In a peer-to-peer model, a device holding a subcollection can synchronize with any partner that holds either the full data collection or a larger subset of the collection. For example, a PDA holding all files in the "/User/Joe/Pictures/SamsBirthday/" directory can synchronize from a laptop storing the "/User/Joe/Pictures/" directory or with the file server storing all of Joe's files. Any replication protocol can be used, but a device's synchronization request needs to explicitly indicate the folder or directory that is being synchronized so the source device can verify that it stores this information and return only items in the specified folder.

Knowledge-driven protocols must address the issue of how to manage knowledge for subcollections. Having each device maintain a single knowledge vector for all items in its stored subcollection may not always be possible. Consider a system with three devices as described above: the file server holds all of Joe's files, the laptop stores files in "/User/Joe/Pictures/," and the PDA stores "/User/Joe/Pictures/SamsBirthday." Suppose that an application directly accessing the file server places a new photo in "/User/Joe/Pictures/Hawaii" with version F6 and also edits a photo stored in "/User/Joe/Pictures/SamsBirthday," resulting in version F7. Now when the PDA synchronizes from the file server, it will receive the edited photo (version F7) but not the new Hawaii photo. How does the PDA update its knowledge? You might think that it could simply update its knowledge vector to indicate that it knows all versions from the file server up to F7; that is alright if the PDA only synchronized with the file server. However, consider what happens when the laptop synchronizes from the PDA. The PDA will send the laptop the updated birthday photo but not the Hawaii photo (since the PDA does not store this photo). If the laptop then updates it knowledge to indicate that it knows up to version F7, it will not be sent the Hawaii photo during its next synchronization with the file server. As suggested above for channel-based systems, one solution is to maintain a separate knowledge vector for each folder in a collection, but this approach quickly becomes unmanageable for file systems with lots of directories. Another approach is to associate knowledge vectors with subcollections. When synchronizing from a partial replica, a device will receive learned knowledge that is tagged with the partner's subcollection. As described in Section 4.3.10, this learned knowledge is added to the device's existing knowledge. This means that devices may need to store multiple knowledge vectors for different subcollections.

5.5   CONTENT-BASED FILTERS

In general, a mobile device with limited storage may specify, via a content-based *filter*, the items that it wishes to receive and retain in its local replica. Think of a filter as an arbitrary query over the contents of a collection (including its metadata). For example, Bob's cell phone may store phone numbers of people in his address book marked as "personal," his e-mail messages that are from his boss and have not yet been read, evening appointments from his wife's calendar, and low-resolution versions of family photos with a five-star rating.

For complex filters, one option is to propagate all updates to all devices and have each device discard the items that do not match its filter. This requires few, if any, changes to the discussed replication protocols and works for both state- and log-based protocols. It also permits devices to use any filtering techniques without standardizing on a filter query language. The drawback, of course, is that network resources are consumed propagating a potentially large number of updates to mobile devices that are not interested in them.

A more network efficient alternative is for a device to communicate its filter to its synchronization partners. This could happen on each synchronization request or only as needed when the filter changes (assuming the partners are willing to retain filter information across synchronization sessions). The process of filtering items can then take place on the sending side of a synchronization protocol, thereby limiting the items sent over the network. A standard language for specifying filters is required as well as universal mechanisms for applying filters. If a filter is a Boolean predicate over a single data item, a device during synchronization can decide whether to send an updated item without consulting other items in the collection.

For device–master state-based replication protocols, filtered synchronization is relatively easy. The metadata required are the same as for full replication, and the only difference is that the master applies a filter to the updated items that it sends. The device, when sending items to the master, does not apply any filters. For systems that use operation logs, filtering is more difficult. In particular, filters need to apply to operations rather than items. In practice, this means that devices can filter updates based on the type of an operation or the items that it updates, which is rarely as effective as content-based filters.

One slight complication arises with content-based filtering, called the *move-out* problem. For a partial replica defined by a filter, the device usually wants to store items that match the filter *and only those items*. If an item that previously did not match the device's filter is updated so that it now matches the filter, the updated item should be sent to the device. That happens automatically in any of the replication protocols. If an item that previously matched a filter is updated so that it no longer matches the filter, then the item should be discarded by the device. This is the tricky part. Filtered synchronization, as described above, would simply ignore such items. Thus, the partial replica would retain the old version of the updated item. One solution is for the device to (at least periodically) inform the master of the complete set of items that it stores. The master can then return special move-out notifications for items that no longer match the device's filter. This technique can also be used when a device changes its filter.

Filtering in the context of peer-to-peer replication protocols is challenging. For one thing, constraints must be placed on the synchronization topology to ensure eventual convergence for full replicas. As an example, consider a system with two full replicas and one partial replica. If the devices with full replicas never synchronize directly but only synchronize through the device with a partial replica, then an item created in one full replica that does not match the partial replica's filter will never propagate to the other full replica, and vice versa. One, but certainly not the only, way to guarantee converge is to ensure that all full replicas have a well-connected synchronization topology and that partial devices only synchronize with full replicas.

Metadata exchange protocols can support content-based filtering and move-outs in a peer-to-peer topology since each synchronization involves sending the partial replica's full metadata.

How to support peer-to-peer replication among filter-defined partial replicas while obtaining the efficiency benefits of a knowledge-driven protocol remains an open issue for future research.

5.6 CONTEXT-BASED FILTERS

An extension of content-based filtering takes into account the context of the device or user when filtering items. A person's context might include his recent and upcoming appointments, business and personal contacts, current location, important tasks, and critical business variables, such as inventory levels. By examining contextual data, a relatively static context-based filter can anticipate the person's changing information needs. For example, a mobile salesperson may want information about each customer that she is visiting this week to be replicated automatically onto her laptop. Items of interest might include customer profiles, outstanding product orders, notes from past meetings, and even the location of nearby coffee houses. Next week, when the salesperson visits a completely different set of customers, she wants different information (without changing her context-based filter).

Although replication protocols supporting context-based filters are similar to those for content-based filters (discussed in the previous section), the fundamental difference is that the devices that apply filters need access to the relevant contextual information. Filtering is based not only on the contents of an individual item but also on data that exists outside the replicated data collection. For instance, in the salesperson example above, the filtering process needs access to the person's calendar (and perhaps their customer relationship database) to determine this week's customer visits.

As an example, Cogenia's Context Server allowed contextual filters to be specified for each mobile device in an organization [71]. These filters ran on a centralized server that stored not only the customer data that was selectively replicated onto devices but also each employee's calendar and location. With context-based filters, the items of interest to a person (and her mobile device) can vary drastically over time, making efficient support for move-outs critical. The Cogenia server maintained a snapshot of each device's replica, recording the items and versions currently stored on the device. Context-based filters were run periodically on the server to determine the items that should be on the device and then incremental updates computed based on the server-maintained snapshots. Thus, when a mobile device connected to the server, the set of updates for the device was already prepared.

* * * * *

CHAPTER 6

Conflict Management

To reduce reliance on network connectivity, minimize response times, and maximize data availability, mobile devices generally allow their users to update locally stored information at any time. Policies for updating local data objects fall into two categories, those that permit conflicting updates and those that avoid conflicts.

Conflict avoidance techniques either require a device to contact most replicas when updating a data object, using algorithms such as weighted voting [21], or to obtain a lock on the object before updating it. Given the unreliable nature of mobile devices and wireless networks, communicating with other replicas when performing updates is problematic, to say the least. In principle, a device could proactively obtain exclusive locks on a set of objects while connected to a lock granting service and later update those objects while disconnected. Unfortunately, this requires a person to anticipate the objects he might want to update in disconnected mode, prevents others from updating such objects even if the lock holder does not actually need all of the locks that it obtained, and necessitates techniques for recovering locks from poorly behaving or permanently disabled devices. Thus, conflict avoidance techniques are rarely used in mobile systems.

Commonly, mobile systems allow both read and write operations to be performed on locally replicated data objects without coordination with other replicas. Concurrent attempts to update one or more data objects may conflict. This chapter explores the nature of such conflicts, presents alternative mechanisms for detecting conflicting updates, and discusses automatic conflict resolution.

6.1   WHAT IS A CONFLICT?

The one common attribute of all conflicts is that they result from concurrent updates. Two updates were performed concurrently if they were made independently without knowledge of each other. Conflicts involving two update operations are called *write/write* conflicts; *read/write* conflicts go undetected in most mobile systems since reads are allowed to return stale data. Even just considering update conflicts, different definitions and types of conflicts are possible.

Many systems only concern themselves with *single-object concurrency conflicts*, that is, concurrent updates to the same data object. In such systems, different objects are treated independently.

A generalization of this conflict model is *multiobject concurrency conflicts*, in which concurrent updates to any of a set of objects are considered to be conflicting. For example, all of the entries in a person's calendar might be considered an object set for purposes of conflict detection. Any independent updates or additions to the calendar would be viewed as potentially conflicting. Generally, such a definition overstates the set of operations that actually conflict.

In the database world, optimistic concurrency control is based on the notion of *transactional conflicts*. A transaction reads some objects and updates some possibly different objects. Two transactions, T1 and T2, are determined to conflict if T1's read-set includes some object that was written by T2 and T2's read-set includes some object written by T1. In other words, the transactions, when executed concurrently, do not produce the same result as their serial execution. If each transaction is restricted to reading and writing a single object, then transactional conflicts are identical to single-item concurrency conflicts.

Some collaborative systems detect *operational conflicts*. Such systems define conflicts using a two-dimensional table in which each row and each column is an operation that can be performed on replicated data objects. Each cell of the table indicates whether the two operations specified by the associated row and column conflict if executed concurrently. For example, in a banking application, concurrent deposit operations never conflict, whereas withdrawals may conflict with other withdrawals since they could cause one's bank account balance to dip below zero.

Finally, a system might support *semantic conflicts* defined by arbitrary constraints between items. The classic database integrity constraint that says an employee cannot make more money than his manager is one example. An update conflict occurs when concurrent operations cause a constraint to be violated, although each operation individually may succeed. File systems, for example, generally require each file in a directory to have a unique name. If a new file is created on one machine and another file created concurrently on different machine with the same name, then those two files' create operations conflict. As another example, consider a calendar application in which two appointments conflict only if they involve a common participant and overlap in time. Such conflicts might arise if two people are simultaneously adding appointments to a shared calendar from different replicas. For this application, single-item concurrency would fail to detect such conflicts, and multi-item concurrency applied to the whole calendar would detect far too many conflicts. Semantic conflicts are also called *application-specific conflicts* since the constraints being enforced depend heavily on the application and types of objects that it manages.

6.2 CONFLICT DETECTION
6.2.1 No Conflict Detection
The simplest approach for dealing with conflicting updates is just to ignore them. This does not, however, imply that the system cannot provide eventual consistency. Suppose, for example, that an update timestamp is included in the metadata associated with each version of an object. When

an object is updated, the updating device reads it local clock to generate a new update timestamp. When a device receives a new version of an object via replication, it compares the new version's timestamp to its currently stored timestamp for this object. Each replica keeps the version with the latest timestamp, and discards older versions. This technique was used in the Grapevine system [9]. Essentially, timestamps provide a global order for all versions of all objects; other ordering mechanisms are possible. Having a global order on updates ensures that replicas eventually converge to a consistent state. The downside is that, when conflicting updates do occur, the update that was performed "last" according to the generated timestamps always "wins" and the other conflicting updates are quietly discarded. If a mobile device has a slow clock, it runs the risk of having its updates disregarded.

6.2.2 Version Histories

To reliably detect all single-object concurrency conflicts, a system may record the complete *version history* for each data object (or set of objects when detecting multiobject concurrency conflicts). Conceptually, a version history is a directed graph in which each node is a version and each edge represents a causal dependency between versions. When an object is updated, its new version is added to its version history with a link from the previous version that was stored by the updating device. During replication, versions are sent with their version histories. Two version histories for the same object can easily be merged. Two versions conflict if they appear in the same version history and no directed path exists from one to the other. This indicates that the versions were produced concurrently.

Version histories are sufficient for detecting concurrency conflicts but have the unfortunate property of growing proportional to an object's update rate. Their unbounded size can be a problem for mobile devices with limited storage. Some systems truncate version histories by removing versions that are older than a threshold, say 1 month, or by keeping a fixed number of the most recent versions. Although this saves space, this can lead to missed or false conflicts.

6.2.3 Previous Versions

Some systems, such as PRACTI [8], maintain the *previous version* of each item in addition to its current version. Think of this as an extremely truncated version history in which only the latest edge is kept. When an item is updated, its current version is stored as the previous version and then a new current version is generated. The replication protocol sends both the current and previous versions of each updated item. Two different versions of an item are judged to be conflicting if they have the same previous version. This clearly indicates that the two versions were produced independently from the same base version.

Unfortunately, this simple scheme can miss conflicts if the replication protocol does not deliver all versions to all replicas, such as in a state-based system. As an example, consider a scenario in which device A updates version V0 of an item producing version A1. Device A, before synchronizing with any other device, updates the same item a second time, producing version A2. Concurrently, device B updates the item, producing a conflicting version B1. When device A receives version B1, it will not detect a conflict since versions B1 (with a previous version of V0) and A2 (with a previous version of A1) have different previous versions. Similarly, device B will not detect a conflict when receiving version A2, although it would have if it first received version A1.

One might be tempted to think that changing the conflict detection algorithm could solve this problem without requiring additional metadata. In particular, suppose that two different versions are deemed conflicting if neither is the other's previous version. In this case, versions A2 and B1 in the previous scenario would be correctly determined to conflict. However, this revised conflict definition can lead to false conflicts. Consider the scenario where device A produces version A1 based on version V0 and then produces version A2. Device B receives version A2 of this item and then updates it, producing version B1. No conflicting updates have been performed. However, suppose that device C stores version A1 for this item with previous version V0 and then receives version B1 with previous version A2. Device C would incorrectly deduce that versions A1 and B1 conflict.

These two scenarios point out the subtle difficulty of detecting conflicts without maintaining complete version histories. Nevertheless, previous versions can be used to reliably detect conflicting updates in some replication protocols. In particular, in a log-based system in which each replica receives all updates and updates are consistently ordered in each replica's log, conflicting versions can be detected as follows. Suppose that update operations with associated versions are applied to each replica in the order in which they appear in the log and that this order is consistent with the order in which these versions were produced. When applying a logged update, a conflict is detected if the current version of the item being updated is not the same as the previous version associated with this update. In the case of concurrent update operations, the first operation that appears in a device's log will succeed, whereas the second operation will be detected as a conflict (since the execution of the first conflicting operation would have changed the version to one that differs from the previous version expected by the second operation).

6.2.4 Version Vectors

Version vectors (as described in Section 4.4.5) were designed for the Locus-replicated file system [64], and it has been shown that they can determine whether any two operations in a distributed system were performed concurrently. Thus, versions vectors are widely used in replication protocols to detect single-object concurrency conflicts. Specifically, a version vector is stored with each item, updated when the item is updated, and propagated along with the item. Given two versions of an item with their associated version vectors, the versions conflict if neither version vector dominates the other.

Essentially, a version vector captures the item's version history in a compact form. Instead of growing proportional to the number of updates, a version vector remains a fixed size that is based on the number of replicas. A new replica need only be added to the version vectors of items that it updates. Retired replicas must be retained in version vectors indefinitely. Thus, systems with dynamically changing replica membership will experience version vectors that grow over time even if the number of active replicas remains relatively constant. For large numbers of replicas and small items, the overhead of storing a version vector for each item may be substantial.

6.2.5   Made-With Knowledge

To detect concurrency conflicts, knowledge-driven protocols have used a scheme that it similar to version vectors. Each version of an item is associated with a knowledge vector called *made-with knowledge*. Recall from Section 4.2.11 that each device maintains a knowledge vector to compactly record the versions that have been incorporated into its replica. When a device updates an item, in addition to producing a new version for the item, it stores its current knowledge as the made-with knowledge for the updated item.

As with versions vectors, made-with knowledge is sufficient for detecting concurrent, and therefore conflicting, operations on individual data items. Suppose that Bob and Alice both update the same item, producing new versions. Bob's version follows Alice's version if and only if Alice's version was included in Bob's knowledge at the time that Bob updated the item, that is, Bob's version was produced on a device that was aware of Alice's version. These two versions conflict if neither follows the other, that is, if both (1) Alice's version is not included in the made-with knowledge of Bob's version and (2) Bob's version is not included in the made-with knowledge of Alice's version. Note that, unlike version vectors, the made-with knowledge of different versions is never directly compared.

Remarkably, the WinFS system showed that per-item made-with knowledge need not be stored explicitly, thereby substantially reducing the amount of persistent metadata [57]. In particular, a replica's base knowledge can be used in place of each item's made-with knowledge. Although the replica's knowledge contains many versions that would not be contained in an item's made-with knowledge, these extra versions do not affect the determination of whether two versions of the same item conflict.

6.2.6   Read-Sets

In systems that support atomic transactions, a device may read and write multiple items in a single transaction. When a transaction commits, a device should record the versions of items read by a transaction along with the updated items (in a state-based system) or logged operations (in a log-based system). The item's *read-set*, consisting of a <item ID, version> pairs, can later be used to detect transaction conflicts [11]. A conflict is detected when, at the time that an updated item or update operation is received and processed by a device, some item in the received read-set has a stored version in the device's replica that differs from the version recorded in the read-set. This implies that a previously received or local update modified some item on which the newly received update depends.

6.2.7   Operation Conflict Tables

For systems that provide operation logging, conflict tables can be defined to indicate which concurrent operations conflict with which other operations. An e-mail application, for example, might provide a conflict table indicating that "mark message as read" never conflicts with "mark message as spam" even if they both update the same item. Conceptually, when receiving a new operation to be performed, a replica can detect conflicts by searching for conflicting operations in its log. Although conceptually clean, conflict tables are not practically useful, in general, for two basic reasons. First, whether two operations conflict often depends not only on the types of the operations but also on the parameters to these operations and perhaps on other state. For example, two debit operations on a bank account conflict only if the total amount being debited exceeds the balance in the account. Second, the cost of conflict detection grows with the size of the operation log since the arrival of a new operation via replication requires finding and checking all concurrent operations in the local log. In practice, this is expensive, and it is not possible unless an infinite log is maintained.

6.2.8   Integrity Constraints

Checking arbitrary integrity constraints can be done by database triggers when an update is performed for both state-based and log-based replication. If a constraint is violated, then the update is determined to conflict with some previously performed update. Application-specific conflicts need only be specified in term of the data invariants that the application is expected to maintain. Such invariants can apply to any number of data items.

   Multi-item constraint checking is possible in state-based replication but is not well-suited to the replication model. The problem is not in checking the constraints. The problem is that state-based replication assumes that data items can be treated independently and, thus, during synchronization, sends updated items in an arbitrary order. Unfortunately, reordering updates can cause a remote replica to detect a constraint violation even for a valid sequence of local operations, thereby resulting in false conflicts. Consider the following scenario. A user moves his dentist appointment from Monday morning to Tuesday afternoon and then adds a haircut appointment for Monday morning. If the new haircut appointment is sent to a remote device before the modified dentist appointment, the device will detect a violation of the constraint that the user cannot get his hair cut and teeth checked at the same time. This problem with constraint checking does not arise for log-based replication because the operation log naturally maintains the order of local operations, and the replication protocol preserves this order.

6.2.9   Dependency Checks

As another approach to detecting semantic conflicts, the Bayou system associated an application-provided *dependency check* with each logged update operation (or transaction). As the term implies, each dependency check is intended to capture the state on which an operation depends and is a pre-condition for the operation's successful execution. In Bayou, a dependency check is represented as query and an expected set of results. Before applying a received update operation, a device first runs the dependency check's query against its local replica and confirms that the returned query results match the expected results. If the expected results are obtained, i.e., the dependency check succeeds, then the update operation is performed. If the dependency check fails, then the operation conflicts with some operation that appeared earlier in the device's log.

Dependency checks are powerful enough to implement a variety of conflict detection tech-niques. For example, the "previous version" scheme can be obtained by having the query in a de-pendency check return the current version of the item being updated and expecting the result to be the updating device's previous version. Read-sets can be emulated in a similar fashion by query-ing the versions of each item in the operation's read-set. To enforce application-specific integrity constraints, the dependency check can query for any items that violate the constraint and expect no results. For example, when creating a new file, a file system could use a dependency check that queries for existing files with the same name and expect no such files to be found. A calendar system could use dependency checks to query for conflicting appointments, i.e., appointments that overlap in time.

The main disadvantage of dependency checks is the cost of storing, transmitting, and execut-ing them for each update operation. Also, application writers must provide the appropriate depen-dency checks when items are updated, but this is an inherent burden in any application-specific conflict detection scheme.

6.3   CONFLICT RESOLUTION

Once a conflict is detected, steps must be taken to *resolve* the conflict. Resolution involves choosing new contents for the item that was updated concurrently and generating a new version that super-sedes previous known versions. The item's new contents could be taken from one of the conflicting versions or produced by merging the conflicting versions in some manner. In any case, when all conflicts have been resolved, the system should be left with a single latest version of each data item. Issues in conflict resolution include how, when, and where to resolve conflicts as well as how to ensure that all replicas eventually agree on the conflict resolution and thereby converge to a consistent state.

6.3.1 How Are Conflicts Resolved?

Mobile systems have generally used one of three main approaches for resolving conflicts. Each of these approaches may be preferred in some environments and may better meet the needs of particular applications. Therefore, some systems provide all three as options.

One, the system can rely on users to choose the "winner" from among two or more conflicting versions of an item, which is referred to as *manual conflict resolution*. When a conflict is detected at device, that device simply records this fact and takes no other action. The device temporarily retains multiple conflicting versions of an item in its local replica or in a special *conflict log*. Sometime later, the conflicting versions are presented to a human, perhaps through a special conflict resolution tool, and the user selects which version should be kept. Of course, if the user is not happy with either of the versions or wishes to produce a merged version, the user can resolve the conflict and then immediately update the item with the desired contents, thereby overwriting the resolution. Thus, this approach is simple and flexible. The main drawback is that it places an added burden on users. In some situations, the device that detects a conflict may not have originated either of the conflicting updates; it merely received them from others. In this case, asking this device's owner to resolve the conflict may be impractical since this person may be unaware of the correct resolution.

Two, a system can have a built-in *conflict resolution policy* or allow system administrators to select from a well-defined set of policies that are applied to each detected conflict. When a conflict is detected, the winning version is chosen without human involvement based on the configured policy. For instance, a "last writer wins" policy may be enforced, in which the conflicting version with the latest update timestamp is retained. Alternatively, if the detected conflict involves an update that was performed locally and one that originated on another device, then a "local update wins" (or perhaps a "remote update wins") policy may be preferred. In a device–master replication model, a "master wins" policy may be chosen. Different conflict resolution policies could be enforced for different types of items. Policy-based conflict resolution ensures that conflicts are resolved in a timely manner, most likely immediately after detection, and requires no human interaction or application extensions. Assuming that devices enforce identical policies, it also ensures that devices will resolve conflicts identically. Unfortunately, mobile devices that wish to replicate data are often owned by different people who administer their own devices independently. Expecting device owners to set uniform policies may be unrealistic.

Three, a system can allow applications to register *conflict resolvers* that are automatically invoked when a conflict is detect. Such resolvers are software routines that are presented with the conflicting versions, allowed to take arbitrary action, and expected to return new contents for the item in conflict. For state-based replication, resolvers can be registered for various types of data items, providing type-specific resolution procedures. As an example, suppose that a person maintains a list of favorite Web sites on both her mobile phone and her laptop. In case of conflicting updates to this person's favorites list, a special resolver could be invoked that merges the conflicting lists while removing duplicate entries rather than simply choosing one version over the other, thereby losing entries. For operation-sending protocols, resolvers can be associated with operation types, thereby ensuring that their actions are consistent with the semantics of the intended operations. For example, Coda provides special built-in resolution mechanisms for handling conflicting directory operations, such as name conflicts. Conflict resolvers allow applications to extend the set of automatic conflict resolution policies provided by a system. They are a powerful mechanism for dealing with conflicts in an application-specific manner but do raise concerns about safety, since resolvers perform arbitrary computations, and about replica convergence as discussed below.

6.3.2   Where Are Conflicts Resolved?

Regardless of the approach taken to resolve conflicting items, a question remains about when and where conflict resolution takes place. In other words, which devices in a system are responsible for performing manual or automatic conflict resolution? There are two basic alternatives: the resolve-everywhere model and resolve-anywhere model.

In the *resolve-everywhere* scheme, conflicting versions of an item fully propagate to all replicas, and each device independently detects conflicts and resolves them locally. Resolutions do not propagate to other devices; they simply affect a device's local replica. This is the approach that has been taken by most replication systems.

This scheme is unsuitable if the system relies on manual conflict resolution since the owners of each device would be asked to independently resolve each conflict, which, besides being a significant annoyance, can lead to owners making inconsistent decisions. Thus, this technique is mainly used when conflicts are resolved automatically.

With automatic conflict resolvers, the main drawback of this approach is that it relies on deterministic conflict resolution at each device to achieve overall replica convergence. This means that each device needs to (1) have the same conflict resolvers installed and (2) execute these resolvers deterministically. Bayou solved the first issue by including conflict resolution code (called *mergeprocs*) with each propagated update [95], which uses extra bandwidth to send the resolvers and assumes that devices are willing to run conflict resolution code that they receive from others. Even so, ensuring deterministic execution among a diverse set of mobile devices can be tricky. Moreover, this approach uses computational resources at each device to run conflict resolvers, and it uses network bandwidth to fully propagate conflicting updates. Thus, it consumes mobile devices' precious energy.

In the *resolve-anywhere* scheme, when a device detects a conflict and resolves it locally, the device also propagates its resolution to other replicas. Conflict resolution produces new updates that are sent via the normal replication protocol and overwrite the conflicting versions. This works well in systems with manual resolution since a human need only resolve a conflict on one device, as well as in systems that combine manual and automatic conflict resolution. In systems with automatic conflict resolvers, it allows certain devices, such as those with excess resources, to be configured as conflict resolution sites, whereas others simply log conflicts until they learn of the chosen resolution. Conflict resolvers need no special privileges; they behave like any other application-level program that reads and writes replicated data. This is the approach taken in WinFS.

The principal advantage of this scheme is that convergence is guaranteed even if different replicas have different conflict resolvers (or if their conflict resolvers are nondeterministic) as long as the replication protocol reliably propagates updates. This approach also potentially reduces the bandwidth usage since conflicting versions need not be sent to all replicas, only their resolution. In particular, if all devices have automatic conflict resolvers, then a conflict is resolved when first detected by any device.

One drawback of this scheme is that conflicts may be concurrently detected at multiple devices that will then introduce concurrent, and hence conflicting, resolutions. Detecting identical updates can eliminate "false" conflicts in many cases but does not completely solve the problem. Consider the case where two devices make conflicting updates to an item. Two other devices, C and D, detect the conflict and run their automatic conflict resolver, which is deterministic and produces the same result at C and D. So, if devices C and D synchronize at this point, then we have no problem. But before synchronizing with anyone, suppose the user at C makes a subsequent update, perhaps because he did not like the result of the automatic resolution. Now, user C's update will be detected as conflicting with the D's resolution, although it should not.

Additionally, concurrent conflict resolution could lead to a "conflict resolution war," that is, an indefinite sequence of conflict resolutions where conflict resolvers try to resolve conflicts produced by conflict resolvers. Consider the following scenario involving devices A, B, C, and D with replicas of some collection. Assume that all of the replicas start out in a mutually consistent state. A performs a local update. B also performs a local update, thereby introducing a conflict. A now syncs with C. B now syncs with D. So far, A and C have the same data, and B and D have the same data. A now syncs with B. At this point, A or B, whichever first receives the other's update, detects a conflict and resolves it. Suppose that A detects the conflict and decides that its local change wins.

After two-way synchronization, A and B will now have A's change. C and D now sync with each other. Suppose that D detects the conflict between A and B's changes and decides that B's change wins. C and D now end up with B's change. Now, A and C sync again, and B and D sync again. Again, they detect conflicts and resolve them differently. This can continue indefinitely without the replicas ever converging to the same state.

To avoid conflict resolution wars, a system can ensure that automatic conflict resolvers make deterministic decisions regardless of where and when they are invoked. This means that even if two devices independently detect and resolve a conflict, they will choose the same version. Alternatively, the system can maintain a flag associated with each version of an item indicating whether this version was produced by a human or by a conflict resolver. If a conflict is detected between two versions produced by conflict resolvers, it is not automatically resolved. Such conflicts can be ignored if the two versions have identical data (and the smallest version can be retained as the current version). If two conflict resolvers produced different data, then their conflicting resolutions can either be resolved according to a deterministic policy, such as "last writer wins," or logged as a conflict that needs to be resolved manually.

* * * *

CHAPTER 7

Case Studies

This chapter presents the design of a number of systems that have been developed to support replication of data among mobile devices. The intent is to show how designers have chosen techniques outlined in previous chapters and integrated them into practical systems with varying characteristics. Some of the systems discussed are research prototypes, whereas others are products that are available today. Table 7.1 compares these systems using the fundamental questions posed at the beginning of Chapter 4.

7.1 CODA
7.1.1 History and Background
The Coda research project at CMU has had a remarkable history of innovation in the mobile computing space. The project started in 1987 as a follow-up to the Andrew File System with the intent of building a highly available file system. Early focus was on replicating files among servers. Soon, the demand for running Coda file system clients on laptops, which were just starting to gain popular use in the research community, forced the Coda project to explore support for disconnected operation; wireless networks were not yet common, and hence, laptops spent a considerable amount of time without any network connectivity. Later, the project investigated weakly connected operation, namely, how to take advantage of low bandwidth connectivity to perform functions such as trickle reintegration. More recently, the project tackled issues such as translucent caching, isolation-only transactions, and operation shipping. As a research project, Coda produced impactful results for more than twelve years, and as a deployed system, Coda is still in use today [86].

7.1.2 Target Applications
Having been developed in a university environment, Coda's main user community consisted of students and faculty. Hence, it supported users editing papers, developing code, producing talk slides, sending e-mail, and so on. In general, any application that reads and writes Unix files could store its data in Coda.

7.1.3 System Model
One of Coda's key design tenets was that authoritative data should reside on servers, which can be locked in machine rooms and maintained to provide a secure computing environment. Clients,

TABLE 7.1: Comparison of select systems

SYSTEM	DATES	DATA	MODEL	CONSISTENCY	PROTOCOL	CONFLICTS
Coda (CMU)	1987–	Files	Client–server	Weak while disconnected, isolation-only transactions	multiRPC to servers, log reintegration after disconnection	Per-file timestamps, automatic conflict resolvers
Ficus, Rumor, Roam (UCLA)	1990–1998	Files	Peer-to-peer	Weak	Best-effort multicast plus state-based anti-entropy	Per-file version vectors, automatic conflict resolvers
Bayou (Xerox PARC)	1992–1997	Databases	Peer-to-peer	Eventual, session guarantees	Pairwise log reconciliation	Per-update application-specific dependency checks and merge procedures
Sybase iAnywhere	~1995–	Databases	Client–server	Weak while disconnected	Timestamp-based row exchange	Per-row previous version, automatic conflict resolvers
Microsoft Sync Framework (MSF)	2001–	XML, databases, files	Peer-to-peer	Eventual	Pairwise state-based reconciliation	Per-replica version vectors, manual or automatic resolution

including mobile devices, who are inherently untrustworthy, store cached copies of files that only they access. Thus, there is a strong client–server model with a rigid division of responsibilities between first-class replicas (on servers) and second-class replicas (on client devices). Files are organized into volumes, which denote subdirectories in the file name space. Each volume can be replicated on a set of stationary servers. Clients cache a subset of the files in a volume, i.e., perform whole-file caching.

While connected to one or more servers, a client device accesses files from the servers and caches them on its local disk. Write operations are performed on all available servers in parallel. Hoarding is used to preload files into the client's cache in anticipation of disconnection. While disconnected, the device's cache serves application read requests, emulating responses that would be received from a server. Write operations, as well as file create and delete and other directory operations, are recorded in a local log on the disconnected device. When communication with the servers is restored, a reintegration process replays the logged operations to bring the servers up-to-date. In summary, each Coda client is in one of three states: hoarding, emulation, or reintegration [45].

7.1.4   Consistency

Coda's semantics might be described as "accessible" consistency. When connected, a client retrieves file metadata from all available servers to ensure that it reads the most recent version of the file; the client's update operations are performed on all available servers, thereby keeping the servers in a mutually consistent state. However, network partitions and server failures may cause server replicas to diverge and clients to access out-of-date file copies. When a partition is restored or a server recovers, clients are responsible for detecting inconsistencies and initiating recovery to bring the servers back into a consistent state. Additionally, servers send callbacks to inform clients when cached files have been updated.

Thus, when all servers and clients are well-connected, Coda provides reasonably strong consistency. However, when failures occur, the system falls back into a weak consistency mode. Clients in different partitions, for instance, may update disjoint sets of servers and observe different file contents. Similarly, when clients become disconnected, either voluntarily or involuntarily, such clients may continue operation by reading stale data, and any updates produced will not be available to others until reintegration occurs.

7.1.5   Replication Mechanisms

Coda provides two basic replication protocols, one for server replication and one for client–server reintegration. During normal (nondisconnected) operation, a client sends operations to the available volume storage group (AVSG) using a special multiRPC protocol. For read operations, servers return their latest "storeid," which the client uses to decide from which server it should retrieve the file contents and also to determine whether the servers in the AVSG are inconsistent and need to be repaired. For write operations, the client assigns a unique storeid consisting of the client's unique identifier and an update counter. Each available server performs the write operation and returns an acknowledgment to the client including a version vector (called the CVV) that indicates the updates it has performed and which updates it knows have been performed by other servers. The client then reports to the AVSG which members successfully completed the write by sending a final CVV. Directory operations are handled in a similar manner.

During reintegration following a period of disconnection, a client sends its entire replay log in parallel to the servers in the AVSG, and servers perform each logged operation just as they would during normal operation, except that all operations are committed in a single transaction [45]. Log compaction is performed so that obsolete log entries can be discarded, such as when a file is written multiple times during disconnected operation or when a file is created and then deleted before reintegration. Once the reintegration process completes, a client frees its replay log. Later versions of Coda included techniques for trickle reintegration, whereby logged operations can be incrementally and asynchronously applied to servers when there is a low-bandwidth, high-delay connection [58].

7.1.6   Conflict Handling

Coda uses version vectors (see Section 6.2.4) to detect conflicting updates performed in different partitions of a volume storage group. To detect conflicts performed by clients while disconnected, the latest storeids recorded at the client and servers are compared during the reintegration process.

For conflicting directory operations, Coda uses built-in resolution procedures based on the semantics of Unix directories (as was proposed in the Locus system) [49]. For resolving write conflicts on files, clients can automatically invoke application-specific resolvers [50, 51]. Such resolvers run on the client and produce a new file content that is written to the servers. Appropriate resolvers are selected based on rules that can include file types and pathnames. For cases where no application-specific resolvers are suitable, Coda provides a repair tool that allows users to carry out manual resolution.

7.2   FICUS
7.2.1   History and Background

The Ficus distributed file system was developed at UCLA starting in the late 1980s [24, 25, 63]. It was an intellectual descendant of the Locus distributed operating system, which included a file system that could tolerate network partitions. Ficus carried this work forward by concentrating on an NFS-compatible file system with a peer-to-peer model, more scalable algorithms than Locus, and additional support for conflict resolution. Motivated by the needs of mobile computing users, the Rumor [26] and then Roam [74] file systems followed Ficus with a similar peer-to-peer architecture. Much of the basic design discussion below applies to Rumor and Roam as well as to Ficus.

7.2.2   Target Applications

Like Coda, Ficus was designed as a Unix file system. Thus, it can store files for any Unix file-based application.

7.2.3   System Model

Ficus uses a peer-to-peer model, and all replicas behave as equals. File volumes can be replicated on any Ficus node. Ficus clients can access replicas on their local disk and can also mount remote file systems that are accessed via the NFS protocol. Early versions of Ficus used whole volume replication. Selective replication was added later so that a device, such as a mobile device with limited disk space, could choose to store an arbitrary subset of the files in a volume [72]. Devices keep track of where files are replicated using the same optimistic replication protocol used to propagate updated files and directories.

7.2.4   Consistency

Clients can read and update any locally or remotely accessible file without coordinating with other devices. In other words, Ficus provides a read-anywhere, update-anywhere model that the developers call *one-copy availability*. It guarantees eventual consistency through pairwise reconciliations.

7.2.5   Replication Mechanisms

Each update operation in Ficus performs the update on a single replica of a file. Other available replicas are notified that the file has been updated using a best-effort messaging protocol. Upon receiving notice that a file has been updated, a replica explicitly "pulls" the new file contents from the replica holding the latest version. It can do this immediately or delay retrieval until a more suitable time or until a later notification arrives for the same file, in which case, the previous notification can be ignored.

Since a device may miss update notifications because of unreliable message delivery or because it was disconnected from the updating device, a directory reconciliation protocol is used to detect missing updates. Reconciliation involves direct communication between pairs of devices. Each device periodically contacts another device that replicates the same file volume(s) and retrieves metadata about each remotely stored file and directory. These metadata are compared with that of locally stored files to determine which files on the remote device are more recent and need to be fetched. Reconciliation is a heavyweight process as it requires a device to potentially examine, communicate, and process information about each file. Timestamps can be used to narrow down the set of files that must be considered during reconciliation.

Although Ficus supports any reconciliation topology, having each of n replicas reconcile with all other replicas would result in n^2 reconciliations. To reduce this cost, replicas typically are arranged in a ring so that each replica has two regular reconciliation partners. The ring topology is adaptable so that disconnected replicas can be removed and later reinserted when they reconnect. In a system with partial replicas using selective replication, a separate ring may be needed for each file since different files can be stored at different sets of replicas [72].

Rumor uses a reconciliation protocol that is similar to that developed for Ficus but skips the best-effort update notification step. That is, Rumor simply relies on periodic reconciliation between replicas to propagate updates (as is done in Bayou). The reasoning is that notification is ineffective for primarily disconnected mobile devices. Roam adopts a similar approach but, for increased scalability, organizes replicas into a two-level "ward" model rather than a single adaptable ring. In the ward model, nearby replicas are grouped into wards, and replicas within a ward reconcile with each other on a regular basis. Less-frequent reconciliation occurs between ward leaders.

To avoid create/delete ambiguities, deleted files are retained by a device in its local replica until the device knows that every other device knows that it is aware of the deletion. This requires a two-phase garbage collection protocol [25].

7.2.6   Conflict Handling

Ficus relies on per-file version vectors to detect update conflicts. Version vectors were invented for Locus and inherited by Ficus, Rumor, and Roam. Concurrent updates to a directory or a given file are detected during reconciliation (or when the file is fetched following an update notification) by comparing the version vectors associated with the directory or file stored on different replicas.

Conflicting directory information is automatically resolved using built-in policies based on the semantics of Unix file directories (as was done in Locus and Coda). For example, current updates to a directory are resolved by merging the directory contents from the conflicting versions.

When a replica detects a conflict on a file, it searches for a locally registered conflict resolver based on the type of the file and the file name. A conflict resolver is a program that takes two conflicting versions of a file and produces new file contents. A number of different conflict resolvers were developed for Ficus, including ones for mail directories, log files, and control files [75]. If a suitable automatic resolver is not found, then the file's owner is sent an e-mail message reporting the conflict and is expected to manually resolve it. When a file conflict is resolved, either automatically or manually, its version vector is updated to dominate those of the previously conflicting versions.

7.3   BAYOU

7.3.1   History and Background

The Bayou system was developed at Xerox PARC between 1982 and 1986 as part of its research program on ubiquitous computing. It borrowed ideas from previous work at PARC on epidemic algorithms while explicitly targeting mobile computing devices with intermittent connectivity.

7.3.2   Target Applications

Bayou's aim was to support interpersonal collaborative applications. Several sample applications were designed and built including a mail reader called BXMH, a calendar sharing application, and a shared bibliographic database.

7.3.3   System Model

Bayou provides a peer-to-peer model in which each mobile device stores a full copy of a shared relational database. Update transactions can be initiated at any device. Periodically, a device chooses another replica from which to pull updates that it does not yet know, i.e., updates propagate via a pairwise synchronization protocol. When a device receives a new update, either from a local application or during synchronization, it adds the update to its log. A device's database is obtained, at least conceptually, by applying all updates in the order that they appear in its log. All devices play identical roles in Bayou except for one designated primary replica, which takes on the additional burden of choosing a final commit order for all updates.

7.3.4   Consistency

The Bayou system was designed to ensure eventual and causal consistency (see Section 3.3) while accommodating arbitrary update operations and communication topologies. It accomplished this through a fault-tolerant log-based delivery mechanism coupled with a means for globally ordering updates. Applications could choose to read tentative data that might be undone later because of conflicting operations or restrict their access to committed results. Additionally, applications could select from four different session guarantees that offer increased consistency (see Section 3.4).

7.3.5   Replication Mechanisms

Bayou used a log-based, knowledge-driven protocol (as discussed in Section 4.3.10). Each update is assigned a version stamp consisting of the device originating the update and an update counter. Each device maintains a version vector to capture the set of versions that has been received, stored in its log, and applied to its database replica. One device initiates reconciliation with another device by sending its version vector. The second device responds by returning any updates stored in its log whose versions are not covered by the received version vector. To preserve causal consistency, updates are sent in the order that they appear in the sending device's log over an order-preserving transport protocol, such as TCP. A device's version vector is incrementally updated as new updates arrive.

To ensure that replicas converge to a mutually consistent state, updates are ordered consistently at all replicas using their version stamps along with commit stamps assigned by the primary replica. Since updates originating at different devices may arrive at other devices in different orders and since updates are applied to a device's replica as soon as they are received, reconciliation may cause previously received updates to be rolled back and reapplied after newly arriving updates. The designated primary replica assigns commit stamps indicating the final order for all updates. Once a device learns the commit stamps for updates in a prefix of its log, it can be certain that these updates will not need to be rolled back. Such updates move from the "tentative" to the "committed" state.

7.3.6 Conflict Handling

Perhaps Bayou's greatest contribution was its development of application-specific techniques for detecting and resolving conflicts. In particular, as discussed in Section 6.2.9, each logged update operation is accompanied by a dependency check used to detect semantically conflicting operations. Dependency checks are more powerful than the per-item version vectors used in many systems since they can detect conflicts involving multiple items. For resolving conflicts, each update operation also included a merge procedure, which was an interpreted code fragment that could read the database and issue a new update based on the current state.

7.4   SYBASE iANYWHERE

7.4.1   History and Background

Sybase was founded as a database company in 1984 and emerged as the leader in mobile database management software in the mid- to late 1990s. Today, its iAnywhere suite of mobile software includes the SQL Anywhere DBMS and the MobiLink synchronization technology. SQL Anywhere is a full-featured relational database management system that runs on a variety of computing platforms from large servers to laptops. For small-footprint mobile devices, such as smartphones, Sybase's UltraLite database runs on Windows Mobile, Palm OS, and Symbian operating systems. MobiLink allows intermittently connected devices to synchronize data with enterprise servers.

7.4.2 Target Applications

Sybase focuses mainly on business applications in which access to enterprise data is critical. Customers include everything from salespeople, who share product catalogs and sales records, to shipping companies that track the locations of goods on trucks, planes, and ships.

7.4.3 System Model

MobiLink allows synchronization between a centralized database server, called the *consolidated database*, and any number of *remote databases* running on fixed or mobile devices. Remote databases must be either SQL Anywhere or UltraLite databases, whereas the server can run any of the major relational database management systems, including SQL Anywhere, Oracle, IBM DB2, or Microsoft SQL Server.

A remote database may be a partial replica of the consolidated database through both horizontal and vertical partitioning. That is, a device may hold a subset of the database tables, rows, or columns. This subset can be defined by an arbitrary SQL query that filters data sent to the device. In some cases, the remote database may have a different database schema altogether, although that requires special schema mapping scripts to be installed on the server.

While disconnected from the server, a device can run applications that perform local transactions to read and write its remote database. The inserted, updated, and deleted rows resulting from such transactions are logged on the device. If a device wishes to discard some rows without having those rows deleted from the server (and all other devices), it can run a delete transaction with logging turned off. Periodically, each device with a remote database contacts the server to upload locally logged transaction results and download updates made by other devices that have been uploaded since its last synchronization.

7.4.4 Consistency

Devices are allowed to read stale data and perform conflicting updates; in other words, the system provides a weak consistency model. However, all transactions are serialized and transactional boundaries are preserved, meaning that all of the writes in a device-local transaction are performed atomically at the server.

7.4.5 Replication Mechanisms

The MobiLink replication protocol is a relatively straightforward example of the timestamp scheme presented in Section 4.3.6. A device sends its entire transaction log to the server over a wired or wireless network connection using TCP or a similar reliable transport mechanism. When the server has received all of the device's updates, it applies them to the consolidated database atomically as one server-side transaction. This server transaction may abort because of interference from other activity on the server database but will be retried until it commits successfully. The server database includes a timestamp column that records when each row was last updated. Deleted rows are not actually deleted from the server; they are simply marked as deleted in the server database using a special status column or recorded in a special shadow table.

To download updates to the device, the server then runs a query to select those rows that have been updated (or deleted) since the device's last synchronization and that match the device's filter. Of course, this data set includes updates made by the device itself that were just uploaded. The server uses the log that it received from the device to filter out any such updates. The remaining batch of updated rows, which were obtained from other remote databases or from transactions run directly against the server database, are then downloaded to the device. The device applies updates that it receives from the server in a single local transaction. This transaction, in addition to modifying the device's database, updates the device's timestamp of when it last synchronized with the server.

The device treats the response from the server as an acknowledgment that its log was successfully uploaded. The device can thus discard its logged updates. The server may request an acknowledgment from the device, but need not. If the data sent from the server is not received by the device, it will be resent during the next synchronization.

7.4.6   Conflict Handling

A conflict occurs when a row has been updated on both the device and the server since the device's last synchronization. Concurrent updates are detected at the server using per-row timestamps. If the "old" timestamp included in rows that are uploaded from a device does not match the last-updated timestamp in the same rows stored on the server, the server detects a conflict. As an option, the server may be configured to perform column-based rather than row-based conflict detection. This requires an update timestamp for each column. Conflict checking is fairly expensive since it must be done for each row before the server can apply new updates. For this reason, some customers choose to run their servers without conflict detection.

7.5   MICROSOFT SYNC FRAMEWORK
7.5.1   History and Background

In November 2007, Microsoft announced a new data replication platform called the Microsoft Sync Framework (MSF). This is a software library intended for use by applications running on both mobile devices and stationary PCs and servers. This product grew out of an earlier effort that started in 2001 to develop a fresh storage system for Microsoft Windows called WinFS. WinFS, in turn, borrowed its basic replication design from the Bayou system while accommodating some additional constraints and adding some innovations [62]. Although WinFS was removed from the Longhorn release of Windows and has no plans to be released as a stand-alone product, its replication protocol lives on in a number of Microsoft products and services. MSF represents a packaging of this technology for use by third-party applications. It is presented here as an example of a modern replication protocol that was designed with mobility in mind.

7.5.2 Target Applications

MFS was designed to support the replication needs of a wide variety of commercial applications. Many of Microsoft's applications manage personal information, such as e-mail, contacts, task lists, calendars, photos, documents, Web favorites, and so on, and most of these run on mobile devices, including laptops, PDAs, and smartphones. While WinFS relied on an XML data model, MFS can replicate arbitrary data from XML objects to files to relational databases through the use of plug-in data providers.

7.5.3 System Model

The MFS architecture is peer to peer. Full or partial replicas of a data collection can reside on any device, and no devices play any special roles. Typically, an application accesses a replica residing on the same machine on which the application runs. Read and write operations are performed as part of transactions that execute locally. Periodically, devices synchronize data with each other using a protocol that resembles the one developed for Bayou. However, rather than maintaining update logs, devices exchange items drawn directly from their data stores. MFS supports arbitrary sync topologies and schedules, that is, any device can synchronize with any other device at any time.

7.5.4 Consistency

As in Bayou, the update-anywhere model used in MSF coupled with its peer-to-peer synchronization presents applications with weakly consistent data. Eventual consistency is assured as long as the graph of device synchronization partnerships is well-connected. No other consistency guarantees are provided by the basic framework, although applications are welcome to add their own consistency mechanisms.

7.5.5 Replication Mechanisms

MSF uses a knowledge-driven protocol (as discussed in Section 4.3.11). Each replica maintains a knowledge vector as a shorthand representation for the set of versions known to the replica. To initiate synchronization with another replica, a device sends its knowledge to its sync partner. The sync partner returns all items in its data store whose latest version is not included in the requestor's knowledge along with metadata used to detect conflicts. The receiving device can write the complete batch of new items in one transaction or in a sequence of smaller transactions. While adding items to its data store, the device atomically updates its knowledge so that those same items will not be resent in future synchronization operations.

As mentioned above, the MSF resembles Bayou in its system model and sync protocol, with several notable differences. MSF does not use a write log. MSF makes no assumptions about the order in which items are sent (or arrive) during synchronization. Thus, any transport mechanism can be used, including unreliable wireless networks. Lost or reordered items may cause a replica to have "exceptions" in its knowledge but do not cause convergence or performance problems. MSF does not require replicas to rollback tentative updates or maintain a global order on updates. Instead, each item is treated as an independent entity, and replicas need only agree on the latest version of each item. Finally, as explained in the next subsection, MSF includes a new scheme for detecting and resolving conflicts.

7.5.6 Conflict Handling

WinFS showed that the knowledge vectors used during synchronization can also be used to reliably detect concurrent, and hence conflicting, updates to each item (see Section 6.2.5) [57]. MSF adopted this same technique for conflict detection. When a conflict is detected at a device, automatic conflict resolvers are invoked, if possible, to resolve the conflict by producing a new version. The new, resolved version of the item then propagates to other replicas via the normal synchronization protocol. If a conflict cannot be resolved locally, then the conflicting versions of an item continue to propagate to other replicas, where they may be automatically or manually resolved. This approach allows conflict resolvers to be installed at some but not all replicas and ensures convergence even if resolvers have nondeterministic executions.

· · · ·

CHAPTER 8

Conclusions

Seamless computing requires that people have ready access to their data at any time from anywhere. In theory, this could be achieved by placing all data in a shared repository, such as a network accessible file server, and having all applications on all devices access this single, shared database. In practice, such a centralized approach to data management is infeasible, except in limited situations, due to nonuniform network connectivity and latencies as well as device limitations and regulatory restrictions. For example, many handheld devices do not have wireless network capabilities, although this is rapidly changing. Even with wireless network adapters, devices may have restricted communication in environments such as airplanes and hospitals. Moreover, the high cost of wide-area wireless networking renders its use less desirable than local data access. Although all of the technology trends are in the right direction, ubiquitous, wide-area, low-cost, high-bandwidth, low-latency network communications are still many years away. Thus, for the foreseeable future, storage systems must provide the ability to replicate data close to its point of use, ideally colocated with applications running on a PC or mobile device.

The same factors that argue for replicating data onto mobile devices dictate a style of replication known as "optimistic" or "update-anywhere" replication in which replicas are allowed to behave autonomously. Specifically, users and applications can read and write data at a single replica without coordinating their activity with other replicas. This provides maximum data availability since the inaccessibility of some remote replicate cannot hinder access to one's locally replicated data. Contrast this to a system in which replicas are kept mutually consistent at all times by using distributed transactions with locking and two-phase commit. In such a system, a slow, failed, or disconnected device negatively affects the overall system behavior. Although the benefits of an update-anywhere replication scheme are evident, the costs and complexities may be subtle. Because replicas diverge through concurrent, autonomous activity, applications must be able to tolerate weakly consistent data. Additionally, to drive replicas toward eventual convergence, mechanisms are needed for disseminating updates and dealing with the potential conflicts that naturally arise in an update-anywhere system.

This lecture presented a variety of replication protocols for propagating updates between devices and techniques for detecting and resolving conflicting updates. These range from simple protocols that use best-effort multicast to complex knowledge-driven protocols that support peer-to-peer delivery over arbitrary connections. Different techniques make widely varying assumptions both about the devices on which they operate and the network characteristics and connectivity of those devices. Replication protocols also differ in the functionality that they provide to mobile applications, such as consistency and delivery guarantees, features that are often apparent to end users. Crafting a mobile application, therefore, requires designers to choose data management technologies carefully and evaluate technology tradeoffs to fully meet the needs of the intended user community. This lecture should help guide system designers in this important endeavor.

• • • •

Bibliography

1. A. Acharya and B. R. Badrinath. Delivering multicast messages in networks with mobile hosts. *Proceedings of the 13th International Conference on Distributed Computing Systems*, Pittsburgh, PA, 1993, pp. 292–299.

2. J. E. Allchin. A suite of robust algorithms for maintaining replicated data using weak consistency conditions. *Proceedings of the Third Symposium on Reliability in Distributed Software and Database Systems*, Clearwater Beach, FL, October 1983, pp. 47–56.

3. R. Alonso, D. Barbara, and H. Garcia-Molina. Data caching issues in an information retrieval system. *ACM Transactions on Database Systems* 15(3):359–384, September 1990.

4. R. Alonso and H. F. Korth. Database system issues in nomadic computing. *Proceedings of the ACM SIGMOD International Conference on Management of Data*, Washington, DC, May 1993, pp. 388–392.

5. R. Alonso, E. Haber, and H. F. Korth. A database interface for mobile computers. *Proceedings of the 1992 Globecom Workshop on Networking of Personal Communication Applications*, December 1992.

6. M. Balazinska, A. Deshpande, M. J. Franklin, P. B. Gibbons, J. Gray, S. Nath, M. Hansen, M. Liebhold, A. Szalay, and V. Tao. Data management in the worldwide sensor web. *IEEE Pervasive Computing*, April–June 2007, pp. 30–40.

7. D. Barbara and H. Garcia-Molina. *Replicated Data Management in Mobile Environments: Anything New Under the Sun?* Department of Computer Science, Stanford University, Stanford, CA, August 29, 1993.

8. N. Belaramani, M. Dahlin, L. Gao, A. Nayate, A. Venkataramani, P. Yalagandula, J. Zheng. PRACTI replication. *Proceedings of the USENIX Symposium on Networked Systems Design and Implementation (NSDI)*, May 2006.

9. A. D. Birrell, R. Levin, M. D. Schroeder, and R. M. Needham. Grapevine: an exercise in distributed computing. *Communications of the ACM* 25(4):260–274, April 1982.

10. L. P. Cox and B. D. Noble. Fast reconciliation in fluid replication. *Proceedings of the International Conference on Distributed Computing Systems (ICDCS)*, April 2001.

11.   S. Davidson, H. Garcia-Molina, and D. Skeen. Consistency in a partitioned network: a survey. *ACM Computing Surveys* 17(3):341–370, September 1985.

12.   A. Demers, D. Greene, C. Hauser, W. Irish, J. Larson, S. Shenker, H. Sturgis, D. Swinehart, and D. Terry. Epidemic algorithms for replicated database maintenance. *Proceedings of the Sixth Symposium on Principles of Distributed Computing*, Vancouver, BC, Canada, August 1987, pp. 1–12. Also in *ACM Operating Systems Review* 22(1):8–32, January 1988.

13.   A. Demers, K. Petersen, M. Spreitzer, D. Terry, M. M. Theimer, and B. Welch. The Bayou architecture: support for data sharing among mobile users. *Proceedings of the Workshop on Mobile Computing Systems and Applications*, Santa Cruz, CA, December 1994. IEEE Press, Piscataway, NJ.

14.   A. Downing. Conflict resolution in symmetric replication. *Proceedings of the European Oracle User Group Conference*, Florence, Italy, April 1995, pp. 167–175.

15.   M. Ebling, L. Mummert, and D. Steere. Overcoming the network bottleneck in mobile computing. *Proceedings of the Workshop on Mobile Computing Systems and Applications*, Santa Cruz, CA, December 1994. IEEE Press, Piscataway, NJ.

16.   W. K. Edwards, E. D. Mynatt, K. Petersen, M. J. Spreitzer, D. B. Terry, and M. M. Theimer. Designing and implementing asynchronous collaborative applications with Bayou. *Proceedings of the Symposium on User Interface Software and Technology (UIST)*, Alberta, Canada, 1997, pp. 119–128.

17.   C. A. Ellis and S. J. Gibbs. Concurrency control in groupware systems. *Proceedings of the ACM SIGMOD International Conference on Management of Data*, Portland, OR, June 1989, pp. 399–407.

18.   K. Fall. A delay-tolerant network architecture for challenged Internets. *Proceedings of the Conference on Applications, Technologies, Architectures, and Protocols for Computer Communications*, Karlsruhe, Germany, 2003, pp. 27–34.

19.   J. Flinn and M. Satyanarayanan. Energy-aware adaptation for mobile applications. *Proceedings of the ACM Symposium on Operating Systems Principles (SOSP)*, Charleston, SC, December 1999, pp. 48–63.

20.   J. Flinn, S. Sinnamohideen, N. Tolia, and M. Satyanarayanan. Data staging on untrusted surrogates. *Proceedings of the 2nd USENIX Conference on File Storage and Technologies (FAST)*, San Francisco, CA, March 2003, pp. 15–28.

21.   D. K. Gifford. Weighted voting for replicated data. *Proceedings of the Symposium on Operating Systems Principles (SOSP)*, Pacific Grove, CA, December 1979, pp. 150–162.

22.   R. A. Golding. A weak-consistency architecture for distributed information services. *Computing Systems* 5(4):379–405, Fall 1992.

23. J. Gray, P. Helland, P. O'Neil, and D. Shasha. The dangers of replication and a solution. *Proceedings of the 1996 ACM SIGMOD International Conference on Management of Data*, June 1996, pp. 173–182.

24. R. G. Guy, J. S. Heidemann, W. Mak, T. W. Page Jr., G. J. Popek, and D. Rothmeier. Implementation of the Ficus replicated file system. *Proceedings of the Summer 1990 USENIX Conference*, Anaheim, CA, June 1990, pp. 63–71.

25. R. G. Guy, G. J. Popek, and T. W. Page Jr. Consistency algorithms for optimistic replication. *Proceedings of the First International Conference on Network Protocols*, October 1993. IEEE Press, Piscataway, NJ.

26. R. Guy, P. Reicher, D. Ratner, M. Gunter, W. Ma, and G. Popek. Rumor: mobile data access through optimistic peer-to-peer replication. *Proceedings of the ER Workshop on Mobile Data Access*, 1998, pp. 254–265.

27. J. S. Heidemann, T. W. Page Jr., R. G. Guy, and G. J. Popek. Primarily disconnected operation: experiences with Ficus. *Proceedings of the Second IEEE Workshop on Management of Replicated Data*, Los Angeles, CA, November 1992, pp. 2–5.

28. J. Holliday, R. C. Steinke, D. Agrawal, and A. El Abbadi. Epidemic algorithms for replicated databases. *IEEE Transactions on Knowledge and Data Engineering*, 15(5):1218–1238, 2003.

29. P. Honeyman, L. Huston, J. Rees, and D. Bachmann. The LITTLE WORK Project. *Proceedings of the Third Workshop on Workstation Operating Systems*, Key Biscayne, FL, April 1992, pp. 11–14.

30. J. H. Howard. Using reconciliation to share files between occasionally connected computers. *Proceedings of the IEEE Workshop on Workstation Operating Systems*, Napa, CA, October 1993, pp. 56–60.

31. Y. Huang, P. Sistla, and O. Wolfson. Data replication for mobile computers. *Proceedings of the ACM SIGMOD International Conference on Management of Data*, Minneapolis, MN, May 1994, pp. 13–24.

32. Y.-W. Huang and P. Yu. Lightweight version vectors for pervasive computing devices. *Proceedings of the 2000 International Workshops on Parallel Processing*, 2000, pp. 43–48. IEEE Press, Piscataway, NJ.

33. L. B. Hutson and P. Honeyman. Disconnected operation for AFS. *Proceedings of the USENIX Mobile and Location-Independent Computing Symposium*, Cambridge, MA, August 1993, pp. 1–10.

34. L. Huston and P. Honeyman. Peephole log optimization. *Proceedings of the Workshop on Mobile Computing Systems and Applications*, Santa Cruz, CA, December 1994. IEEE Press, Piscataway, NJ.

35. T. Imielinski and B. R. Badrinath. Querying in highly mobile distributed environments. *Proceedings of the International Conference on Very Large Databases (VLDB)*, Vancouver, BC, Canada, 1992, pp. 41–52.

36. T. Imielinski and B. R. Badrinath. Data management for mobile computing. *SIGMOD Record* 22(1):34–39, March 1993.

37. T. Imielinski and B. R. Badrinath. Mobile wireless computing: challenges in data management. *Communications of the ACM* 37(10):18–28, October 1994.

38. A. Imine, P. Molli, G. Oster, and M. Rusinowitch. Achieving convergence with operational transformation in distributed groupware systems. INRIA Technical Report No. 5188, May 2004.

39. S. Jain, M. Demmer, R. Patra, and K. Fall. Using redundancy to cope with failures in a delay tolerant network. *Proceedings of the ACM SIGCOMM*, Philadelphia, PA, 2005, pp. 109–120.

40. A. D. Joseph, J. A. Tauber, and M. F. Kaashoek. Mobile computing with the Rover toolkit. *IEEE Transactions on Computers* 46(3):337–352, March 1997.

41. A. D. Joseph and M. F. Kaashoek. Building reliable mobile-aware applications using the Rover toolkit. *Wireless Networks* 3(5):405–419, October 1997.

42. R. Katz. Adaptation and mobility in wireless information systems. *IEEE Personal Communications Magazine* 1(1): 6–17, 1995.

43. A.-M. Kermarrec, A. Rowstron, M. Shapiro, and P. Druschel. The IceCube approach to the reconciliation of diverging replicas. *Proceedings of the 20th annual ACM SIGACT-SIGOPS Symposium on Principles of Distributed Computing (PODC '01)*, Newport, RI, August 26–29, 2001.

44. M. Kim, L. P. Cox, and B. D. Noble. Safety, visibility, and performance in a wide-area file system. *Proceedings of the Conference on File and Storage Technologies (FAST)*, January 2002.

45. J. J. Kistler and M. Satyanarayanan. Disconnected operation in the Coda file system. *ACM Transactions on Computer Systems* 10(1):3–25, February 1992.

46. N. Krishnakumar and A. J. Bernstein. Bounded ignorance in replicated systems. *Proceedings of the Symposium on Principles of Database Systems*, 1991, pp. 63–74.

47. G. H. Kuenning. The design of the SEER predictive caching system. *Proceedings of the Workshop on Mobile Computing Systems and Applications*, Santa Cruz, CA, December 1994. IEEE Press, Piscataway, NJ.

48. G. Kuenning and G. J. Popek. Automated hoarding for mobile computers. *Proceedings of the Sixteen ACM Symposium on Operating Systems Principles*, Saint Malo, France, October 1997, pp. 264–275.

49. P. Kumar and M. Satyanarayanan. Log-based directory resolution in the Coda file system. *Proceedings of the Second International Conference on Parallel and Distributed Information Systems*, San Diego, CA, January 1993.

50. P. Kumar and M. Satyanarayanan. Supporting application-specific resolution in an optimistically replicated file system. *Proceedings of the IEEE Workshop on Workstation Operating Systems*, Napa, CA, October 1993, pp. 66–70.

51. P. Kumar and M. Satyanarayanan. Flexible and safe resolution of file conflicts. *Proceedings of the USENIX Winter 1995 Conference on Unix and Advanced Computing Systems*, New Orleans, LA, January 1995, pp. 16–20.

52. R. Ladin, B. Liskov, L. Shrira, and S. Ghemawat. Providing high availability using lazy replication. *ACM Transaction on Computer Systems*, 10(4):360, November 1992.

53. L. Lamport. Time, clocks, and the ordering of events in a distributed system. *Communications of the ACM* 21(7):558–565, July 1978.

54. Y.-W. Lee, K.-S. Leung, and M. Satyanarayanan. Operation-based update propagation in a mobile file system. *Proceedings of the 1999 USENIX Technical Conference*, Monterey, CA, June 1999.

55. T. Liu, C. M. Sadler, P. Zhang, and M. Martonosi. Implementing software on resource-constrained mobile sensors: experiences with Impala and ZebraNet. *Proceedings of the 2nd International Conference on Mobile Systems, Applications, and Services (MobiSys)*, Boston, MA, June 2004, pp. 256–269.

56. Y. P. Lum and F. C. M. Lau. User-centric content negotiation for effective adaptation service in mobile computing. *IEEE Transactions on Software Engineering* 29(12), pp. 1100–1111, December 2003.

57. D. Malkhi and D. Terry. Concise version vectors in WinFS. *Distributed Computing* 20(3):209–219, October 2007.

58. L. Mummert, M. Ebling, and M. Satyanarayanan. Exploiting weak connectivity for mobile file access. *Proceedings of the 15th ACM Symposium on Operating Systems Principles*, Copper Mountain, CO, December 1995, pp. 143–155.

59. D. A. Nichols, P. Curtis, M. Dixon, and J. Lamping. High-latency, low-bandwidth windowing in the Jupiter collaboration system. *Proceedings of the ACM Conference on User Interface System Technology (UIST)*, Pittsburgh, PA, 1995, pp. 111–120,

60. B. D. Noble, M. Satyanarayanan, D. Narayanan, J. E. Tilton, J. Flinn, and K. R. Walker. Agile application-aware adaptation for mobility. *Proceedings of the Sixteenth ACM Symposium on Operating Systems Principles*, Saint Malo, France, October 1997.

61. B. D. Noble, B. Fleis, and L. P. Cox. Deferring trust in fluid replication. *Proceedings of the 9th ACM SIGOPS European Workshop*, Kolding, Denmark, 2000, pp. 79–84.

62. L. Novik, I. Hudis, D. B. Terry, S. Anand, V. Jhaveri, A. Shah, and Y. Wu. Peer-to-peer replication in WinFS. *Microsoft Research Technical Report* MSR-TR-2006-78, June 2006.

63. T. W. Page Jr., R. G. Guy, J. S. Heidemann, D. H. Ratner, P. L. Reiher, A. Goel, G. H. Kuenning, and G. Popek. Perspectives on optimistically replicated peer-to-peer filing. *Software—Practice and Experience* 28(2):155–180, February 1998.

64. D. S. Parker Jr., G. J. Popek, G. Rudisin, A. Stoughton, B. J. Walker, E. Walton, J. M. Chow, D. E., S. Kiser, and C. Kline. Detection of mutual inconsistency in distributed systems. *IEEE Transactions on Software Engineering* 9(3):240–247, May 1983.

65. K. Petersen, M. J. Spreitzer, D. B. Terry, M. M. Theimer, and A. J. Demers. Flexible update propagation for weakly consistent replication. *Proceedings of the of the 16th ACM Symposium on Operating Systems Principles* (SOSP-16), Saint Malo, France, October 1997.

66. S. H. Phatak and B. R. Badrinath. Transaction-centric reconciliation in disconnected databases. *ACM Monet Journal*, 53, 1999.

67. S. H. Phatak and B. R. Badrinath. Conflict resolution and reconciliation in disconnected databases. *Proceedings of the MDDS 1999*, September 1999.

68. E. Pitoura and B. Bhargava. Building information systems for mobile environments. *Proceedings of 3rd International Conference on Information and Knowledge Management*, 1994, pp. 371–378.

69. E. Pitoura and B. K. Bhargava. Maintaining consistency of data in mobile distributed environments. *Proceedings of the International Conference on Distributed Computing Systems*, 1995, pp. 404–413.

70. R. Prakash, M. Raynal, and M. Singhal. An efficient causal ordering algorithm for mobile computing environments. *Proceedings of the International Conference on Distributed Computing Systems*, 1996, pp. 744–751.

71. M. Rangan, E. Swierk, and D. B. Terry. Contextual replication for mobile users. *Proceedings IEEE International Conference on Mobile Business (ICMB)*, Sydney, Australia, July 2005.

72. D. Ratner, G. J. Popek, P. Reiher, and R. Guy. Peer replication with selective control. *Proceedings of the MDA '99, First International Conference on Mobile Data Access*, Hong Kong, December 1999.

73. D. Ratner, P. L. Reiher, G. J. Popek, and G. H. Kuenning. Replication requirements in mobile environments. *Mobile Networks and Applications* 6(6):525–533, 2001.

74. D. Ratner, P. Reiher, and G. J. Popek. Roam: a scalable replication system for mobility. *Mobile Networks and Applications* 9:537–544, 2004.

75. P. Reiher, J. Heidemann, D. Ratner, G. Skinner, and G. Popek. Resolving file conflicts in the Ficus file system. *Proceedings of the Summer USENIX Conference*, June 1994, pp. 183–195.

76. P. Reiher, J. Popek, M. Gunter, J. Salomone, and D. Ratner. Peer-to-peer reconciliation based replication for mobile computers. *Proceedings of the European Conference on Object Oriented Programming (ECOOP) Workshop on Mobility and Replication*, June 1996.

77. Q. Ren and M. H. Dunham. Using semantic caching to manage location dependent data in mobile computing. *Sixth Annual International Conference on Mobile Computing and Networking (MobiCom'00)*, August 2000.

78. M. Rennhackkamp. Mobile database replication. *DBMS Online* 10(11), October 1997. http://www.dbmsmag.com/9710d17.html.

79. M. Ressel, D. Nitsche-Ruhland, and R. Gunzenhauser. An integrating, transformation-oriented approach to concurrency control and undo in group editors. *Proceedings of the ACM Conference on Computer Supported Cooperative Work*, Cambridge, MA, 1996, pp. 288–297.

80. Y. Saito and M. Shapiro. Optimistic replication. *ACM Computing Surveys* 37(1), 2005.

81. S. K. Sarin and N. A. Lynch. Discarding obsolete information in a replicated database system. *IEEE Transactions on Software Engineering* SE-13(1):39–47, January 1987.

82. M. Satyanarayanan. Mobile computing. "Hot Topics" column. *IEEE Computer*, pp. 81–82, September 1993.

83. M. Satyanarayanan. Mobile information access. *IEEE Personal Communications* 3(1): 26–33, 1996.

84. M. Satyanarayanan, J. J. Kistler, P. Kumar, M. E. Okasaki, E. H. Siegel, and D. C. Steere. Coda: a highly available file system for a distributed workstation environment. *IEEE Transactions on Computers* 39(4):447–459, 1990.

85. M. Satyanarayanan, J. J. Kistler, L. B. Mummert, M. R. Ebling, P. Kumar, and Q. Lu. Experience with disconnected operation in a mobile environment. *Proceedings of the USENIX Mobile and Location-Independent Computing Symposium*, Cambridge, MA, August 1993, pp. 11–28.

86. M. Satyanarayanan. The evolution of Coda. *ACM Transactions on Computer Systems* 20(2):85–124, May 2002.

87. B. N. Schilit, N. L. Adams, R. Want. Context-aware computing applications. *Proceedings of the Workshop on Mobile Computing Systems and Applications*, Santa Cruz, CA, December 1994. IEEE Press, Piscataway, NJ.

88. M. D. Schroeder, A. D. Birrell, and R. M. Needham. Experience with Grapevine: the growth of a distributed system. *ACM Transactions on Computer Systems* 2(1):3–23, February 1984.

89. R. Schwarz and F. Mattern. Detecting causal relationships in distributed computations: in search of the holy grail. *Distributed Computing* 7(3):149–174, 1994.

90. M. Spreitzer, M. Theimer, K. Petersen, A. J. Demers, and D. B. Terry. Dealing with server corruption in weakly consistent, replicated data systems. *Proceedings of the ACM Conference on Mobile Computing and Networking* (MobiCom), 1997, pp. 234–240.

91. Sybase. *Designing Mobile Applications: Why Sync Is Central*. Sybase iAnywhere Whitepaper. http://www.sybase.com/files/White_Papers/ias_wp_why_sync_is_central.pdf.

92. C. D. Tait and D. Duchamp. Service interface and replica management algorithm for mobile file system clients. *Proceedings of the First International Conference on Parallel and Distributed Information Systems*, December 1991, pp. 190–197.

93. C. D. Tait, H. Lei, S. Acharya, and H. Chang. Intelligent file hoarding for mobile computers. *Proceedings of the ACM Conference on Mobile Computing and Networking (MobiCom)*, Berkeley, CA, November 1995, pp. 119–125.

94. D. Terry, A. Demers, K. Petersen, M. Spreitzer, M. Theimer, and B. Welch. Session guarantees for weakly consistent replicated data. *Proceedings of the International Conference on Parallel and Distributed Information Systems (PDIS)*, September 1994, pp. 140–149.

95. D. B. Terry, M. M. Theimer, K. Petersen, A. J. Demers, M. J. Spreitzer, and C. H. Hauser. Managing update conflicts in Bayou, a weakly connected replicated storage system. *Proceedings of the ACM Symposium on Operating Systems Principles*, Copper Mountain Resort, CO, December 1995.

96. M. M. Theimer, A. Demers, K. Petersen, M. Spreitzer, D. Terry, and B. Welch. Dealing with tentative data values in disconnected work groups. *Proceedings of the Workshop on Mobile Computing Systems and Applications*, Santa Cruz, CA, December 1994. IEEE Press, Piscataway, NJ.

97. N. Tolia, J. Harkes, M. Kozuch, and M. Satyanarayanan. Integrating portable and distributed storage. *Proceedings of the 3rd USENIX Conference on File and Storage Technologies (FAST)*, San Francisco, CA, March 2004, pp. 227–238.

98. J. Van Der Merwe, D. Dawoud, and S. McDonald. A survey on peer-to-peer key management for mobile ad-hoc networks. *Computing Surveys* 39(1):1–45, April 2007.

99. G. M. Voelker and B. N. Bershad. Mobisaic: an information system for a mobile wireless computing environment. *Proceedings of the IEEE Workshop on Mobile Computing Systems and Applications*, Santa Cruz, CA, 1994.

100. B. Walker, G. Popek, R. English, C. Kline, and G. Thiel. The LOCUS distributed operating system. *Proceedings of the Ninth ACM Symposium on Operating Systems Principles*, Bretton Woods, NH, 1983, pp. 49–70.

101. M. H. Wong and D. Agrawal. Tolerating bounded inconsistency for increasing concurrency in database systems. *Proceedings of the Symposium on Principles of Database Systems*, San Diego, CA, June 1992, pp. 236–245.

102. S.-Y. Wu and C.-S. Chang. An active database framework for adaptive mobile data access. *Proceedings of the ER Workshops*, 1998, pp. 335–346.

103. G. Wuu and A. Bernstein. Efficient solutions to the replicated log and dictionary problems. *Proceedings of the Third ACM Symposium on Principles of Distributed Computing*, August 1984, pp. 233–242.

104. H. Yu and A. Vahdat. Design and evaluation of a continuous consistency model for replicated services. *Proceedings of the USENIX Symposium on Operating Systems Design and Implementation*, October 2000, pp. 305–318.

105. H. Yu and A. Vahdat. Efficient numerical error bounding for replicated network services. *Proceedings of the 26th International Conference on Very Large Databases (VLDB)*, September 2000.

106. H. Yu and A. Vahdat. Design and evaluation of a conit-based continuous consistency model for replicated services. *ACM Transactions on Computer Systems* 20(3):239–282, August 2002.

107. M. Weiser. Some computer science issues in ubiquitous computing. *Communications of the ACM* 36(7):75–84, July 1993.

Author Biography

Douglas B. Terry is a principal researcher at the Microsoft Research Silicon Valley laboratory. His research focuses on the design and implementation of novel distributed systems and addresses issues such as information management, fault-tolerance, and mobility. He currently is serving as chair of ACM's Special Interest Group on Operating Systems (SIGOPS). Before joining Microsoft, Doug was the cofounder and CTO of Cogenia, chief scientist of the Computer Science Laboratory at Xerox PARC, and an adjunct professor in the Computer Science Division at UC Berkeley, where he regularly teaches a graduate course on distributed systems. Doug has a PhD in computer science from UC Berkeley.